AF621047

Lecture Notes in Morphogenesis

Series Editor

Alessandro Sarti, CAMS Center for Mathematics, CNRS-EHESS, Paris, France

Advisory Editors

Henri Berestycki, Ecole des hautes etudes en sciences sociales, Paris, France

Paul Bourgine, Ecole Polytechnique, Palaiseau, France

Giovanna Citti, Dipartimento di Matematica, Università di Bologna, Bologna, Bologna, Italy

Paolo Fabbri, Universitá di Venezia, Venzia, Italy

Vincenzo Fano, Department of Basic Sciences and Foundations, Urbino University, Urbino, Italy

Sara Franceschelli, ENS, Lyon, France

Maurizio Gribaudi, EHESS, Paris, France

Annick Lesne, LPTMC UMR 7600, CNRS UMR 7600 case courrier 121, Paris cedex 05, France

Giuseppe Longo, Centre Cavaillès, Ecole Normale Supérieure, Paris, France

Thomas Lorenz, Efiport GmbH, Frankfurt School Verlag, Frankfurt am Main, Hessen, Germany

Jean-Pierre Nadal, Service Facturier, EHESS, Paris, France

Nadine Peyriéras, Institut de Neurobiologie Alfred Fessard, CNRS-N&D, Gif sur Yvette, France

Jean Petitot, EHESS, Paris, France

Jan-Philip Schmidt, University of Heidelberg, Heidelberg, Germany

Carlos Sonnenschein, Department of Integrated Physio and Pathobiology, Tufts University School of Medicine, Boston, MA, USA

Ana Soto, Department of IPP, Tufts University, Boston, MA, USA

Federico Vercellone, University of Turin, Turin, Italy

Lecture Notes in Morphogenesis is an interdisciplinary book series which aims to face the questions of emergence, individuation and becoming of forms from several different points of view: those of pure and applied mathematics, of computational algorithms, of biology, of neurophysiology, of cognitive and social structures. The set of questions above concerns all the manifestations of Being, all the manifestations of Life. At the heart of contemporary embryogenesis lies an essential question: How can form emerge from the constant, chaotic flow? How can a sequence of purely informational elements—an a-signifying combination of chemical substances organized in the DNA molecule—evolve into the highly complex and structured forms of the living organism? A similar question can be asked when we deal with the morphogenesis of vision in neural systems and with the creation of evolving synthetic images, since digital technology makes possible the simulation of emergent processes both of living bodies and of visual forms. Finally the very idea that abstract structures of meaning could be captured in terms of morphodynamic evolution opens the door to new models of semiolinguistics, semiotic morphodynamics, and cognitive grammars. An entire heritage of ideas and concepts has to be reconsidered in order to face new and challenging problems: the theoretical framework opened by Goethe with the introduction of the word "Morphogenesis" is developed by D'Arcy Thompson in "On Growth and Form", it is reorganized with new theoretical insights by the classical structuralism of Levi-Strauss and formalized by the dynamical structuralism of René Thom. The introduction of the post-structuralists ideas of individuation (in Gilbert Simondon and Gilles Deleuze) and plasticity of structures builds a bridge to contemporary problems of morphogenesis at a physical, biological, social and transindividual level. The objective of this book series is to provide suitable theoretical and practical tools for describing evolutionary phenomena at the level of Free boundary problems in Mathematics, Embryogenesis, Image Evolution in Visual Perception, Visual Models of Morphogenesis, Neuromathematics, Autonomy and Self-Organization, Morphogenetic Emergence and Individuation, Theoretical Biology, Cognitive Morphodynamics, Cities Evolution, Semiotics, Subjectivation processes, Social movements as well as new frontiers of Aesthetics. To submit a proposal or request further information, please use the PDF Proposal Form or contact directly: *Dr. Thomas Ditzinger (thomas.ditzinger@springer.com)*

Isabel Marcos · Clément Morier
Editors

The Relevance of René Thom

The Morphological Dimension in Today's Sciences

Springer

Editors
Isabel Marcos
Centre for Philosophy of Science (CFCUL)
University of Lisbon
Lisbon, Portugal

Clément Morier
Jean Moulin University Lyon 3
Lyon, France

ISSN 2195-1934 ISSN 2195-1942 (electronic)
Lecture Notes in Morphogenesis
ISBN 978-3-031-54981-6 ISBN 978-3-031-54982-3 (eBook)
https://doi.org/10.1007/978-3-031-54982-3

This work has been financed by national funds through FCT—Fundação para a Ciência e Tecnologia I.P.—within the framework of the R&D Unit Centre for Philosophy of Science of the University of Lisbon (CFCUL), a strategic project with the FCT IP. Reference: UIDB 00678/2020 and UIDP/00678/2020.

© The Editor(s) (if applicable) and The Author(s), under exclusive license to Springer Nature Switzerland AG 2024

This work is subject to copyright. All rights are solely and exclusively licensed by the Publisher, whether the whole or part of the material is concerned, specifically the rights of translation, reprinting, reuse of illustrations, recitation, broadcasting, reproduction on microfilms or in any other physical way, and transmission or information storage and retrieval, electronic adaptation, computer software, or by similar or dissimilar methodology now known or hereafter developed.
The use of general descriptive names, registered names, trademarks, service marks, etc. in this publication does not imply, even in the absence of a specific statement, that such names are exempt from the relevant protective laws and regulations and therefore free for general use.
The publisher, the authors and the editors are safe to assume that the advice and information in this book are believed to be true and accurate at the date of publication. Neither the publisher nor the authors or the editors give a warranty, expressed or implied, with respect to the material contained herein or for any errors or omissions that may have been made. The publisher remains neutral with regard to jurisdictional claims in published maps and institutional affiliations.

This Springer imprint is published by the registered company Springer Nature Switzerland AG
The registered company address is: Gewerbestrasse 11, 6330 Cham, Switzerland

If disposing of this product, please recycle the paper.

In memory of our professors: René Thom,
Per Aage Brandt, and Jacques Viret

Contents

Chapter 1
What if Science Thought? The Relevance of René Thom or Introducing Topology into Today's Sciences

Clément Morier and Isabel Marcos

1.1 The Relevance of René Thom: A Morphological Look at Neo-Liberal and Logico-Combinatorial Formalism

> It is also the reason why one gear in the watch does not produce another; still less does one watch produce other watches, [by] using (and organizing) other matter for this [production]. It is also the reason why if parts are removed from the watch, it does not replace them on its own; nor, if parts were missing from it when it was first built, does it compensate for this [lack] by having the other parts help out, let alone repair itself on its own when out of order: yet all of this we can expect organized nature to do. Hence an organized being is not a mere machine. For a machine has only motive force. But an organized being has within it formative force, and a formative force that this being imparts to the kinds of matter that lack it (thereby organizing them). This force is therefore a formative force that propagates itself-a force that a mere ability [of one thing] to move [another] (i.e., mechanism) cannot explain.
>
> Kant, *Critique of Judgment*, § 65 [7]

> There was once Someone who could watch the same show [...]
>
> sometimes as a painter would have looked at it, and sometimes as a naturalist;
>
> sometimes like a physicist, and other times like a poet;
>
> and none of these looks were superficial
>
> Paul Valéry, Preface to the Notebooks of Leonardo da Vinci [3]

This work was funded by national funds through FCT—Fundação para a Ciência e a Tecnologia, I.P., within the scope of the R&D Unit Center for Philosophy of Sciences of the University of Lisbon (CFCUL), a strategic project with references FCT I.P. UIDB/00678/2020 and UIDP/00678/2020.

C. Morier · I. Marcos (✉)
Center for Philosophy of Sciences of the University of Lisbon (CFCUL), Lisbon, Portugal
e-mail: p900292@ulusofona.pt

© The Author(s), under exclusive license to Springer Nature Switzerland AG 2024

I. Marcos and C. Morier (eds.), *The Relevance of René Thom*, Lecture Notes in Morphogenesis, https://doi.org/10.1007/978-3-031-54982-3_1

"To Predict is not to Explain". Such was the warning of one of 20th-century France's major thinkers, whose legacy is becoming increasingly inaudible in today's scientific community: René Thom (Fields Medal 1958). To predict, or the ambition of predictability, requires the (potentially infinite) reduction of the forms of nature into discontinuous, quantifiable elements: primarily, so as to enable the subsequent control of their interactions, within a space, according to a mathematical law.

While this logical, combinatorial and quantitative reduction has triumphed in the sciences since the revolution jointly initiated by Descartes and Galileo, it is now extending its reach into contemporary society.

In economics, Marx's deciphering of "Capital" transforms any qualitative form, object or human, into a simple quantity of money, whose speed of propagation on a market generates a tendentially infinite surplus value.

In the way they operate, political societies no longer know how not to define themselves according to an unconscious model of a "market society", linking by law discontinuous elements (individuals) who are equally free, and oriented by economic interests, arranged simply by means of "governance" [5].

This is to say nothing of the extension, into all spheres of social life, of computer control, heir to Cartesian metaphysics, or the reduction of all discourse, speech and practice to undifferentiated quantities of information whose propagation and permutation becomes unlimited by means of networking. Even anthropological and sociological disciplines are endangered by the epistemology of networks, emergences and neuronal combinatorics instead of any human-social specificity [9].

Against such a fiery backdrop, we would like to propose another way of looking at things, one that apprehends forms and restores to natural morphologies the sensitivity of their deployment, when they enter into metamorphosis. A critical view, then, of the neo-liberal reductionist flattening that is spreading through the forms of contemporary rationality. In contrast, a rationalism of forms, of their dynamic of transformation, supported by a vibrant geometry (topology), constitutes an available corpus that restores unity to reality, and avoids its disintegration under a simple shower of atoms [10].

1.2 The Relevance of René Thom: A Shift in Philosophical Perspective

In the History of Ideas, the notion behind the Thomian corpus—carried out from topology, precisely on the basis of the scientific achievements of modernity, and not from a Husserlian "epoché" of natural objectivity [6]—constitutes a rediscovery of the *morphological unthinking* of modern reason, rooted in Aristotle's natural philosophy—in a physics of *ousia* (substance) or "semiophysics", and not in the phenomenological experiences of consciousness.

Thom's scientific discoveries obviously accompanied the phenomenological rehabilitation of sensitive morphologies, but his fruitfulness lay in the production of an explanation and a generativity, by means of the "folding" of space. By integrating into physical objectivity itself the topological and dynamic principles of its phenomenalization—the forms of which could only be apprehended, *de jure*, on the subjective-relative plane for the idealism of mathematical physics, since the Cartesian rupture—Thom joins directly the Notebooks of Leonardo da Vinci, and, with him, the tradition of European Humanism…

By listening to René Thom's lessons, followed by several schools of linguistics, semiotics and theoretical biology, the phenomenalization of physical objectivity can be treated as a new type of natural phenomenon. The task then becomes to integrate into the objectivity of forms the principles of their phenomenalization that carry the layer, or the strata, of meaning.

1.3 The Relevance of René Thom: An Issue About Rationality in Modernity

This book is based on the scientific results of the "Actualité de René Thom" project (2017–2023 (see also [8])). Through interventions—on the use of morphological modelling—or philosophical and epistemological discussions—on the status of René Thom's discoveries—this project has brought to light issues in the history of ideas, around an *aporia*, which is the subject of incessant doctrinal conflicts, throughout scientific modernity:

- Either this modern rationality wants to "rehabilitate" the ancient knowledge of quality, stemming from Aristotelianism—*substances* or *forma substantialis*, interacting with *substrates*, and phenomenalized in natural forms, according to a *realism* of efficient causality. This intellectual arc, from Aristotle to Thom's work, runs through such thinking that of Thomas Aquinas, summoned in the line of Leibniz[1] [14], Goethe, Geoffroy St-Hilaire, D'Arcy Thompson (who began translating the biological works of Aristotle in 1910), A. Turing and C. Waddington;
- Or that mathematical physics—in its fight against the occult qualities so decried by Descartes—proposes to serialize every state of matter in a space of discontinuous points, whose temporal consecution we control, by means of the formalism of a new mathesis. The study of forms would have to give way to the mathematizable legalization of discontinuous entities, offering calculability and predictability by means of functions.

Amid this, the scientific results of the "Actualité de René Thom" project rehabilitate the value, for today's sciences, of a morphological unthought, in relation

[1] See his letter to Des Bosses, dated February 2, 1706: "[…] Therefore, in addition to figure, size and movement, we must admit forms by means of which the difference in appearances arises in matter, forms that can only be intelligibly sought, it seems to me, from *entelechies*." (Frémont [4], p. 85, our translation).

to a rationality built on the idealism of mathematical physics—an idealism that is even accompanied by epistemic locks, which ultimately "foreclose" this unthought, within the great conceptual systems of modernity, such as Kant or Husserl [13].

Many of the results, materializing this morphodynamic rationality at work, construct a history of ideas far removed from a continuous and cumulative development of scientific rationality, based solely on the principle of (efficient) causality. What is to be made of the principle of "structural stability" as a principle of sufficient reason, at least as a foundation for the scientific community? To answer this question is to reposition at its heart the question of the qualitative stability of forms, and that of stable forms of transformation, as integral parts of "reason" in modernity.

This book is part of this modern rationality, morphologically rethought. It does so in two ways: firstly, through its context of enunciation, that of the physico-mathematical formalization of morphological objectivity, within the scientific models of the 20th century; secondly, through René Thom's connection to the Aristotelian doctrine of forms, and to the Aristotelian formula that seems decisive, in Thom's eyes: *è entelechia chorizei*, the act separates. An astonishing insight into the properties of discontinuities "bursting" into a substrate space, resulting in a symmetry-breaking effect, that is as *morphogenic* as it is intrinsically *semiotic*. Homeomeric tissue becomes anhomomeric—formalizes Aristotle, as a biologist—and gives rise to morphogenetic differentiation—recalls Thom, as a topologist…

1.4 The Relevance of René Thom: The Challenges of His Thesis of "Involuted" Meaning in *Phusis* [11]

Physics legalizes phenomena by dealing with laws. The determinism of post-Galilean mechanics, expressed mathematically, substitutes a "generative *logos*" for an explanation by propagation of efficient causality. The geometrization of physical theories trivializes phenomena, reducing their being to a change in spatial arrangement, or to the operativity of groups. The result is an idealist derealization of physical objectivity, or *phusis*. Yet one of the great scientific adventures of the last century was the *Morphological Turn*, which renewed natural philosophy. This is an opportunity to make science rethink, against Heidegger's 1952 formula, through a demand for scientific intelligibility, which is not separated from appearance, nor from a linguistic description of the world [15].

In his mathematical and philosophical prowess, Thom strove to interpret Aristotle as closely as possible to the corpus. As Thom celebrates his centenary (1923–2023), we would like to pay tribute to him, by briefly reviewing three entries from Aristotle's corpus, which show Thom's vision of morphological relevance to the sciences of his time… and by extensions, of our time.

1. **"The limit is also the formal substance of each thing and its quiddity"—Metaphysics, Δ 17, 1022a** [2]: Thom addresses the morphological moment at work in Aristotle, through the concept of "limit", which decides how a form

sends itself into its substrate. Thom's mathematical genius—which intervenes through a peripheral knowledge, topology, a priori far from the issues structuring the history of philosophy—exposes the dynamic modalities by which, precisely, a form is bordered and folded while its contours evolve in space. The foundation of his topological knowledge lies in the singularity, the organizing point of "catastrophe" that emerges abruptly, through a continuous deformation of the substrate. This differentiation, which is morphogenetic, operates through a system of boundaries that organizes the substrate, limits it and breaks down its homogeneity.

2. **"È entéléchia chôrizei", the act separates—Metaphysics, Z 13, 1039a**: Aristotle's proposition guides Thom's neo-Aristotelian epistemology. It is the subject of a specific treatment in his *Semiophysics* ([15], p. 155, p. 227; [15]). Contrary to what phenomenological epistemology assumes, the foundation is not the experience of consciousness, organized as a transcendental system, but the boundary, the catastrophe that divides spaces. The structuring irruption of the finite through morphological differentiation would be the primary condition of all form, phenomenologically perceptible and linguistically describable [12]. "*è entéléchéia chôrizei*", the act of separating, becomes the starting point of a new reading of Aristotle for Thom, valid in modernity: a difference in continuity produces an event of separation, and every morphologically structured substrate derives from this, by deploying the organizational germs of algebraic topology patterns.
3. **"Nature is always in a substrate"—Physics, II, 1, 192b-193a** [1]: Accused of Pythagorean or even Platonic formalism, the sensitivity of Thom's discoveries was to apprehend, not so much the mathematical organization of reality, but rather movement and life, not reducible to logical-combinatorial formalisms, notably present in the computer metaphor of the genetic "programme".[2] For Thom, Aristotle's originality lies in having imposed a new reality at the centre of natural events, which he calls substance (*ousia*), a separable entity closed in on itself, that we can refer to as "*tode ti choriston*". Now, its core, its power of underlying, which makes it not a Pythagorean point or plane, is its substrate (*hypokeimeinon*). Thom understands form in terms of its engagement within a substrate. Thomian morphogeneses do not result from ideal systems, otherwise they would fall victim to a Pythagoreanism incapable of supporting life. He sees forms as a set of qualitative singularities *within* a substrate, never independent of the energetic field—or the dynamical system (of Gradient)—in which these singularities appear.

[2] In Book 14, N, of the Metaphysics, Aristotle discusses the incoherence of a mathematics of numbers to apprehend life and generation in nature. We find extensions of these positions in D'Arcy Thompson, with his theory of Forms, in Goethe, a morphologist attentive to the quarrel between Cuvier and St-Hilaire, and in Leibniz, with his rehabilitation of *forma substantialis*.

1.5 The Relevance of René Thom: A Plurality of Morphological Perspectives Engaged in Their Disciplines

This book is the fruit of collaboration between researchers working on a common platform. It is a collective work, based on a single morphological theme, which aims to capture all the relevance of René Thom's thinking. For each discipline represented here, this morphological theme is approached according to specific, albeit congruent, issues… Here are the main ideas behind the works presented here.

One of the two co-authors of this presentation, **Clément Morier**, demonstrates the importance of René Thom by putting his dynamic model to work, through a dialogue with the philosopher Gilbert Simondon. Far from being relegated to the dustbin of the History of Science, the "morphologies of processes" of catastrophe theory have a specifically innovative aspect: they intrinsically take charge of the reality of movement, the reality of the "time" parameter insofar as it is a change of state and a transformation of form. By studying the morphology known as "swallowtail morphology", and according to the extensions discerned by Thom's last pupil, Jacques Viret (1943–2018)—one of our mentors—the form of an evolutionary process appears topologically. At a time when substantialist and relationalist ontology are at odds, and since *metamorphosis* must be considered in today's sciences, there is nothing that cannot be attempted, not even a reading that draws on the vocabulary of Simondon—a thinker of individuation according to movement—to explain the deployment of René Thom's figures—as the guardian of a pantheon of archetypal morphological deformations.

"We semantize physical space" is the injunction that governed the work of our late friend and other mentor, Pr. **Per Aage Brandt** (1944–2021). His research in semiotics, in particular, demonstrates the relevance of René Thom's morphological thinking. It deconstructs semiotic acts, showing that they function according to a logic of "places", which organize lived spaces, and are thus invested with meaning. His research programme led to what he called a "pheno-physics", after the expression "semiophysics" used by René Thom, or that of "physics of meaning", put forward by our other friend and mentor, Pr. Jean Petitot.

By semantizing, we valorize or repel, we thus invest lived space with repulsive borders or "guardian" edges. Places, that are thus energized, welcome things… and the displacement from their initial places creates narratives of meaning. Brandt gives us the keys to the topological "formants" that establish this *sense-making—mise en sens*—of lived reality, which is a real *morphological shaping—mise en forme*—of the world…

The next study is proposed by **Wolfgang Wildgen**, a long-standing companion of a linguistic science apprehended according to morphological and dynamic "archetypes". Indeed, the author himself belongs to this heritage of formalizing signs and language processes, which is rooted in an entire history of science, the main moments and turning points of which are recounted here. The topicality here lies in the practice of "interdisciplinary research enterprises based on modern mathematics and using the methodological and empirical standards of the natural sciences", to apprehend semantic spaces in general, and linguistic spaces in particular. So, is there an underlying morphology, an archetypal process, that could organize meaningful interactions as diverse as "Charles eats soup, Albert tells a story, Berthe calculates the result"? W. Wildgen's challenge is to plunge us into the heart of a topological and dynamic formalization of such language patterns, even proposing to extend this understanding in terms of dynamic figures to visual and musical signs, bringing René Thom's thinking up to date. Thus, the fundamental activity of the scientist consists in "finding the relevant categories that enable him to grasp the phenomena in his field of observation". So, Thom was not just a mathematician…

The questioning of the relevance of Thom's theses then continues in the fields of geography and urban science: work on the morphogenesis of cities has given rise to uses of René Thom's thinking, while notions dear to the topologist such as deployment or attractors also raise important questions in the discipline of geography, where *space* and *form* are structuring concepts. However, polysemy should not be misleading, and **Olivier Bonin** sets out to establish operative distinctions, to reflect on the expectations of such tools and concepts, before proposing two models applied to urban form, identifying typical parameters and deformations. In this respect, he points out that "the immense enthusiasm for catastrophe theory has given way to a certain scepticism, which has probably prevented a methodical search for applications in certain disciplines such as geography." In his study, he draws on Thom's topological tools, Gilles Ritchot's concepts of structural geography and even George Dumezil's trifunctionality, to help identify several "typical" urban deployments, between which boundaries correspond to changes in urban morphology.

When the book "Morphogenèse de Paris" was published by G. Desmarais, almost 30 years ago, Jean Petitot did not shrink from the significance of such a study, which goes to the root of the discipline of geography: "But its importance to the understanding of the geographical essence of human settlement is such that it merits attention and consideration". Under the active constraint of a geometry of positions, the genesis of the morphological structures of the human-social settlement might spatialize anthropological-symbolic values, whose list of organizing singularities could, with an adequate axiomatic, be inventoried, in order to study their dynamic profile…

However, from the 1990s to the 2010s, the directions actually taken by French-speaking geographical research were unambiguous, according to **Jean-Paul Hubert**: "ideas about the autonomy of forms as theoretical objects (with geometric properties that have a concrete effect on geographical space by overdetermining the mechanisms at work) were consigned to oblivion." Hence the need for Hubert—and this is the whole point of the study he presents—to take stock, 30 years on, of the conceptual motifs, theoretical devices and spin-offs produced in the field of urban form theory, and their modelling—Thom's approach here being perfectly capable of being complemented and/or renewed by others, such as Turing, in particular, insofar as not only "any occupation of space excludes all other occupancies and therefore creates discontinuities", as J-P. Hubert reminds us, but also because urban settlements contain "enabling morphogenes" that self-catalyze through slow and fast dynamics. This is an area of current research that looks set to continue.

The succeeding study marks a transition to the next Part of this book. Through a dialogue between two disciplines, psychoanalysis and semiotics—semiotics being considered here from René Thom's point of view, i.e., dynamically and topologically—**Carlos Farate and Isabel Marcos** propose a co-authored, eminently morphological study. For, this study chooses as the conceptual driving force behind this interdisciplinary dialogue, the notion of *boundaries*. For obvious reasons, since it is by limits that an entity is bordered and its forms can evolve in space. Its morphogenesis, strictly speaking, takes place according to a *system of boundaries*, where differentiations, both structural and functional, operate through a system of *thresholds*. This system of thresholds enables entities to organize themselves on different hierarchical levels, each level having its own stability and points of bifurcation. Here, I. Marcos and C. Farate study the morphogenesis of the structuring boundaries of a territory, based on an analysis of the urban territory, but also of the psychic territory of a human being. The two "territories"—city and psyche—respond to each other in their deployment, providing an "original experience of the limit", as Walter Benjamin might put it.

Boundaries, demarcations, enclaves and thresholds organize the viewpoint of the two researchers, enabling them to carry out a topography of sensitive zones, where singularities emerge and organize the topology of urban or psychic functioning, caught up in systems of attraction and repulsion.

The imaginary of rejection that permeates the urban fabric, and the experience of psychic trauma in an individual, are both linked in an intersubjective relationship "between inhabited territory and the subject that inhabits it". Psychoanalysis and urban semiotics are thus rethought, thanks to an analysis of morphogenetic fields. This is why the two authors attempt to sketch out the contours of what they call territorial psychoanalysis, based on a list of key notions such as *morphogenetic gradient*, *vacuum*, *frontier*, as well as *pregnance* and *salience*. In this sense, the Form is indeed independent of its substrate…

As a result, this book—the fruit of collective research into René Thom's thought, his topology and the epistemological and philosophical consequences of his work—could not completely leave out biology, understood here in the broadest sense of the term, i.e., everything that involves living organisms and is part of evolution. In this sense, the next part had to address such fruitful questions as the uses of modelling in biology; to explore the epistemology of certain concepts, such as the "epigenetic landscape"; while the third part also had to include in its wake topical dialogues in the fields of medicine and psychology—that of both Freud and Jung.

This is why Thom's relevance and the extensions of catastrophe theory thoroughly question the notion of "archetype", a complex notion if ever there was one. Much has been said about this notion, which is mainly deployed in the field of Carl Gustav Jung's analytical psychology: could it be that Thom and Jung—had they been able to talk together—would have discussed their uses of this term, their convergences and divergences, but also the possibility, thanks to the former, of topologically representing dynamic and energetic reality, the effectivity of the archetype, in the psychic evolution that the latter uncovered? This is the challenge that occupied Professor and general practitioner **Jacques Viret**, one of Thom's last pupils, for many years, as he provided extensive extensions to catastrophe theory, making it accessible and revealing the evolutionary continuity that lies at its heart. It is up to the reader to judge whether he has succeeded…

The next study, which pursues Freudian psychoanalysis, is this time situated as much on the epistemological level as on the actual empirical level of the course of therapy: its author, **Benoît Virole** examines the possibilities for catastrophe theory to become a concrete aid for an analyst in a session, understood as including the specificity of a clinical relationship—an intersubjective relationship, tied up with the patient's psyche in a transference relationship', in which the ego passes through varied emotional states, possibly metastable, sometimes on the verge of very critical zones… "As close as possible to the catastrophic singularity, there is a reactive sensitivity that can induce a bifurcation. One could then imagine the art of analytic interpretation as being that of detecting these critical zones, perhaps by intimate resonance between the mental attractors of the analyst and those of his patient." However, this is not the only relevance that the Theory of René Thom offers to psychoanalysis: Thom has shown that actantial schemas exist beneath all subjective representations, organizing stable interactions—*linguistically describable and phenomenologically perceptible*, to use J. Petitot's expression—between the actants/attractors of the process. Now, B. Virole launches the following conjecture, which deserves particular attention as it is so pregnant with future developments: "one could conjecture that the whole of the development of the libido follows the unfolding of a multidimensional aggregative catastrophe of which the stages (oral, anal and phallic) would only be local sections and of which one would have knowledge by their indexical traces, and precisely by the inscriptions of their agent schemes in the analytic material." The symbolic productions of the unconscious would then be part of a *Morphogenesis of the Psyche* whose unfolding we could understand, filled with episodes of regression, surge and turbulence…

This book goes on to highlight the impact of Thom's ideas on the question of modelling in biology. This study is by a specialist in morphogenesis and complexity in biology, Prof. **Jacques Demongeot**, a friend and supporter of René Thom, for whom he was one of the key relays for the application of his ideas and tools in the field of theoretical biology, a field that J. Demongeot notably helped to establish and considerably expand within the scientific community of his time… For him, Thom "traced scientific leads that we faithfully followed and which turned out to be in fact real programmes." This is true, for example, for the notion of an "attractor", which, even though it had already been potently developed by Thom, became an axial notion for studying the processes of morphogenesis and regulation in biology. J. Demongeot clarifies and illustrates this concept to show its impact on the study of stabilization mechanisms in living organisms subject to disturbance. Going back to Buffon, the notion of the attractor would even irrigate the ideas of Lamarck and Claude Bernard, emphasizes Demongeot—without even mentioning the famous 1830 controversy between Cuvier and St-Hilaire about the unity of plan and composition in the animal kingdom, which Goethe followed closely, passionately, and whose reactivation seemed essential to Thom for thinking out a "theoretical" biology, based on his discoveries in topology. Demongeot has promulgated these tools, and the concepts they contain. In this way, he enables us to look back at what is at stake in modelling, in the study of tree growth, for example, or the growth of embryos… "The history of biological modelling shows that, for 250 years, concepts have preceded equations, and that simulations have only confirmed experimental observations, themselves sometimes preceded by rational speculations." By recalling the memory and the work of his friend Jacques Viret, J. Demongeot concludes: "to model the biological or psychological nature, in an almost identical approach, constitutes, since the Age of Enlightenment, the heart of scientific thought."

The next study takes us into a more specific biological field, medicine. As a neurologist, **Pierre Bounolleau** illustrates the importance of Thom's thinking through examples drawn from both "scientific" medicine—which is institutional, causalist, mechanistic and deterministic—and "traditional" medicine, which is diverse but centuries older. Our author restores the scope of so-called "morphodynamic" models (Petitot), i.e. those designed to explain the dynamic processes of morpho-*genesis*—that genesis of structural patterns, based on the sudden appearance of discontinuities, of "singularities", within an underlying progressive and continuous development; these singularities constitute organizing "germs" that initiate a deployment, i.e., once again, the development of a structurally stable—*robust*—evolving process… This type of reflection constitutes an "entire programme" in medicine, where the "constraints of physiological complexity" reign. "The tradition of mathematical formalization is historically limited there" confides P. Bounolleau. This field thus enables us to test the potential of morphodynamic models, because, according to our author, medical progress has come up against "the insurmountable difficulties of modelling for complex phenomena such as cognition and language. In fact, the physical matter of such complex phenomena is neither sensorially observable nor empirically accessible." Thus, through transdisciplinarity, traditional medicine—which could be illustrated by Lévi-Strauss's accounts of cures and the *symbolic efficacy* of certain ancestral practices—and scientific medicine—through the use of connectionist models of cognitive neuroscience, for example—will be able to reconnect… and Thom's relevance will then consist in playing such a *mediating* role!

The exploration of Thom's special relationship with biology continues with a plunge into the history of science and the epistemology of concepts proposed by **Sara Franceschelli.** This constitutes an incursion into the heart of reciprocal influences and borrowings, as well as the ways in which the works of each were received, between Waddington, Thom, Delbrück and Turing—not forgetting the "obligatory" passage via D'Arcy Thompson's masterpiece… Let no one ignorant of *On Growth and Form* enter here! For S. Franceschelli, understanding the epistemological context in which the concept of "structural stability" was applied to the field of biology, enables us to grasp the ways in which decisive notions such as "chreods", "epigenetic landscape" and the "canalization" of a dynamical system have been constructed. For Thom and Waddington, they determine the way we formally conceive the morphogenesis of living organisms, by grasping the internal physics of the process at play and its deployment, "channelled" towards an attractor.

Structural stability is the "conceptual tool", as S. Franceschelli puts it, which enabled Thom to renew the questioning of theoretical biology. In fact, so-called "degenerate" singularities contain information—particularly structural information, which can give rise to functional potential—that can be "generated", precisely, by lifting this initial "degeneracy", using certain parameters of which Thom produced a finite list. These parameters become active in the topological-dynamic functioning of catastrophe theory's figures. These parameters "set in motion" the initial topology and allow it to deploy itself in a process, which is unstable in time, but remains stable—*robust*—at the evolutionary level: we thus obtain the morphology of a process... Biology is thus the preferred field in which to find Thom's signature of catastrophes, even in experimental curves—which is what J. Viret showed in his experiments, to the joy and surprise of his topologist master!

Finally, the book concludes with historical and biographical anecdotes by **Claude P. Bruter**, one of René Thom's close travelling companions, whom Thom met very early on, and who was a key figure in the genesis and development of his ideas. Bruter gives us an insider's view of the famous "Thom seminar" at IHES, in which he was also an early participant. The two scholars forged a friendship based on respect, intellectual exchanges, mutual enrichment and even borrowings. These exchanges made it possible to think as deeply as possible about the decisive notion of "stability", at the heart of catastrophe theory, but also about an epistemology that strives to understand the dynamic conditions under which discontinuities appear within the stability of entities subject to disturbance...

To sum up, may this collective work bring as much intellectual stimulation to its readers as the authors of this introduction took in organizing this research project. Over these 7 years of work, this has been an incomparable intellectual adventure, which has opened up so many crucial avenues of work, for our sciences today... We now leave it all up to the reader.

References

1. Aristote, Physique. Flammarion, Paris (2002)
2. Aristote, Métaphysique. Tome 1, Livres A à Z, Tome 2, Livres H à N. Vrin, Paris (2000)
3. Da Vinci, L.: Préface de Paul Valéry, (1re éd. 1942). Gallimard, Carnets, Paris (2019)
4. Frémont, C.: L'être et la relation. Lettres de Leibniz à Des Bosses. Vrin, Paris (1981)
5. Gauchet, M.: La condition politique. Gallimard, Paris (2005)
6. Husserl, E.: Idées directrices pour une phénoménologie pure et une philosophie phénoménologique. Gallimard, Paris (1985)
7. Kant, I.: Critique of Judgment. Hackett Publishing Company, Indianapolis (1987)
8. Marcos, I., Morier, C.: Centenaire de René Thom (1923–2023): hommage sémiotique et morphodynamique. Estudos Semioticos **19**(1), i–xxviii (2023)
9. Morier, C.: Les morphologies du politique. Approche comparée des œuvres de René Thom et Marcel Gauchet. LGDJ, Paris (2018)
10. Morier, C., Pinchard, B.: René Thom et la réhabilitation des formes substantielles. Philos. Think. Nat. Today **47**, 125–140 (2016)

11. Petitot, J.: Jean Largeault et René Thom: de l'idéalisme du rationalisme physique au réalisme de la philosophie de la nature. In: Espinoza, M. (ed.) De la science à la philosophie. Hommage à Jean Largeault, pp. 47–60. L'Harmattan, Paris (2001)
12. Petitot, J.: Le troisième labyrinthe: dynamique des formes et architectonique disjunctive. In: Berlioz, D., Nef, F. (dir.) L'actualité de Leibniz: les deux labyrinths, pp. 617–632. Franz Steiner, Stuttgart (1999a)
13. Petitot, J.: La généalogie morphologique du structuralisme. Critique **55**(621), 97–122 (1999b)
14. Robinet, A.: Architectonique disjonctive, automates systémiques et idéalité transcendantale dans l'œuvre de Leibniz. Vrin, Paris (1986)
15. Thom, R.: Esquisse d'une sémiophysique. Physique aristotélicienne et Théorie des catastrophes. InterÉditions, Paris (1988)

Part I
René Thom Between Philosophy and Semiotics

Chapter 2
The Form as a Morphology of Process: The Relevance of René Thom for Understanding Individuation

Clément Morier

Thinking about notions such as figure, form, morphology of process, evolving process... leads us to encountering one of the major yet atypical French thinkers of the 20th century, for whom the figure and the form were key words behind any serious explanation concerning the most varied fields of human knowledge. The idea of bringing together science and philosophy—when it comes to examining the explanatory value associated with the handling of certain "archetypical" figures—is one of the singular characteristics of René Thom's work.[1]

However, an aporia occurs at once and constantly arises amidst disputes concerning Thom's discoveries: does not focusing entirely our scientific attention on a singular means of explanation, through forms, lock us up at immediately in the sole framework of *spatiality*? How to give the right importance to genesis and development, whose effects of change seem to require a means of explanation linked more to *temporality*, seen as a metamorphosis, rather than to space? To sum up, is it possible to detect the effects of time through the principles of a form without openly abandoning the stability and solidity which it gives to its object?

Through this aporia, we can grasp the striking issues underlying the problem of configuration and the dynamic organisation of the elements contained in an autonomous unity. René Thom's contribution to these delicate controversies becomes more profound if it enters into discussion with the thought of Gilbert Simondon (1924–1989). This is because he confronted *form* with the terms of its *genesis* and

[1] My warmest thanks and eternal gratitude go to my Mentor Jacques Viret and his memory (1943–2018), for his constant help during my incursion into Thom topology as well as for helping understanding and drawing these figures. *This study is dedicated to him.* It is a reworked and translated version of an earlier work published in *Du mot au concept: Figure*. Grenoble, PUG, 2015. Finally, this text was translated into English by Ian Monk, Member of the Oulipo. Our warmest thanks to him too.

C. Morier (✉)
Jean Moulin Lyon 3 University, Lyon, France
e-mail: clementmorier@yahoo.fr

© The Author(s), under exclusive license to Springer Nature Switzerland AG 2024

I. Marcos and C. Morier (eds.), *The Relevance of René Thom*, Lecture Notes in Morphogenesis, https://doi.org/10.1007/978-3-031-54982-3_2

individuation. We shall attend to this point during our analysis. The implications that can be derived from this will then lead to an enlightenment of the handling of forms inherited from Thom. Finally, by leaving behind the initially phenomenal and static aspect of these forms, our approach will attempt to grasp them through their dispositions so as to describe the internal process of a development. They then take part in the understanding of a development by means of a morphological transformation.

But firstly, the intellectual setting, around which many points attached to the concept of the figure gravitate, will provide this word with its full scientific weight.

2.1 The Form: The Issue of a "*Morphological Turn*"

In his thought about forms, or morphological thought, René Thom (1923–2002), the French mathematician and topologist constantly made his own the famous adage of Sect. 2.11 of Leibniz's *Discourse on Metaphysics*: giving "*post liminio*" back its intellectual legitimacy to reasoning in terms of substantial forms, in direct line with the considerations of Aristotelian qualitative physics [9].

This reasoning has taken several turns: sometimes a stubborn defence of determinism against the reigning "chaos-logic" of the day, noise theory and emergence as the sole source of any possible genesis of natural morphologies [18]; or else a refusal of the Galileo-Newton separation or schism which splits a physical objectivity, reduced to quantity and motion, from the experiential world, or the phenomenal appearance of material forms [14, 15]; or else the classification of phenomena of discontinuity, or breaks in symmetry, by plunging them into continuous surfaces with a greater number of dimensions, thus bringing out a strictly morphogenic organisational aspect.

This final point gave rise to a finite catalogue of figures through which Thom, thanks to his mathematical virtuosity (Fields Medal in 1958) explained phenomenologically the dynamic arrival of the forms of nature: after the advent of Catastrophe Theory the time had come to replace a Wittgensteinian "*Linguistic Turn*" by a "*Morphological Turn*" [16] thus giving back to meaning, language and structures their full referential scope. But a drawback soon arose which cramped the reception of Thom's work in the humanities and social sciences: the great difficulty coming from the algebraic formalism of his topological research, which makes the dryness and density of Thom polynomials almost inaccessible to non-mathematicians.

A departure point enlightens what is at stake in this formalism. A figure, or configuration, can be understood as being related mainly and essentially to space, so to speak. Spatialisation would then be a theoretical inheritance that initiates any modernist scientific legitimacy: Descartes's *res extensa*, or extended substance, then becomes the means to obtain, by using an appropriate point of reference, an integral computability of natural bodies reduced to masspoints in a space whose interactions we control, through quantifiable parameters, by means of a mathematical function [7]. As a good Bourbakist, Thom stood up for this mathematical technicity, but he tried to undermine that mania in the sciences for quantification by producing continuous geometric forms that preside over the qualitative organisation of discrete elements: it is exactly here that the *figure* appears on the stage and provides an explicative value that deserves to be examined more deeply.

Hence Thom's effort to rehabilitate the form as a legitimate means of scientific explanation capable of meeting the issues raised by organisation and regulation, when quantification should be treated with caution.

Alongside his geometrical research, Thom gradually came to formulate a working hypothesis which led to two new insights. Firstly, the following hypothesis: it is necessary to change the research perspective when faced by the difficulties raised by questions of organisation, regulation and the maintenance of the stability of phenomena. Any scientific understanding derived from a Galileo-Newtonian modernity could be faulted in this respect. Physical objectivity is attached to reducing the phenomenal forms that appear to an observer to the combined interactions of the smaller, discontinuous elements that constitute them. A revelatory remark expressed by a major player in this morphologic watershed, C. P. Bruter, presented this problem in its full extension:

> In the analysis of the properties of physical continuums, that of breakability [the quality of being able to be cut] has played an important role. It has led to defining the reductionist point of view. From everything of which a part can be attained, the pure reductionist claims to be able to reconstruct the whole from its elements.
>
> This is to forget that the passage from the whole to a part is accompanied by the destruction of internal tensions, all the bonds that provide the coherency of the edifice, of which a reducer only considers certain homogenous parts. The pure concrete reductionist would affirm that the fashioning of perfectly adapted stones was enough for the construction of magnificent pyramids, which can be taken down and put back up as desired; but he would be forgetting to mention something, firstly the respective positions of the stones in the pyramid, and secondly the weight that allows for the mutual adherence of the pieces of the puzzle. We can thus imagine the effort that would be required to reconstruct the pyramid from blocks that have been crushed and ground into sand. Putting the parts back together would only lead to the reconstitution of a concrete whole if the structural rules pertaining to the nature of this concrete whole were respected, only if it were possible to restore the play of the forces of interaction which provided the stability of the whole ([2], p. 67)

Hence, there are two theoretic ways of proceeding:

(i) It is possible to draw up a list of stable figures which, when associated with basic dynamics, express the form of quantity in regulatory phenomena.
(ii) Thom then viewed the study of processes that mean being able to move from one form to another: he also drew up figures of this. This stability gives them a generalisable value independent from the nature of the forces and the number of interactions at play in the phenomenon to be explained. Topology, or the geometry of elastically deformable surfaces, provides a means of abstraction capable of focusing on the dynamic and energetic schema of an internal means of functioning, which Thom called a "generative *logos*"—in French, a "*logos* générateur" [25]. These figures aim to fix something which is unstable over time (a process) but which are themselves stable in evolutive terms.

It would seem necessary to temporarily abandon a *quantitative* base so as to favour a more *hermeneutic approach* and thus penetrate more deeply into the Heraclitan notion of *meaning*, to which we are invited by the Delphic oracle who neither reveals nor conceals: "ὁ ἄναξ οὗ τὸ μαντεῖόν ἐστι τὸ ἐν Δελφοῖς οὔτε λέγει οὔτε κρύπτει ἀλλὰ σημαίνει"—*The Lord whose oracle is at Delphi neither reveals nor conceals, he gives a sign* (Heraclitus, fragment n° 93).

There then came in Thom's followers a concern to put back *meaning* in what were still-abstract processes, so as to pinpoint their properties by means of a dialogue with concrete domains. This can be seen in semantics [13], embryology and physiology [23, 27, 31] psychology [28, 30] and sociology [10–12]. After over forty years of Catastrophe Theory's being used, the following remark might be made: these forms, which were initially understood as models, were admittedly not exempt from some kind of subsequent confirmation which tautologically repeats experimentally confirmed knowledge. The point was to describe in Catastrophe Theory language processes already set down in natural language, which led to nothing more than a geometric analogy: was this the unescapable step of the first validation of Thom's discoveries' generality? Little by little, the deepening of Catastrophe Theory carried out by Thom's last student, Jacques Viret, gave it a fresh surge, by focusing on the energetic aspect of these forms: he attempted to enlighten qualitatively the plastic retractions pertaining to the manifestation of transformative episodes of an evolving object.

If it is possible, as should be recognised, not to follow Thom in the radicality of the metaphysical demands that drove his scientific arguments, he should still be recognised as having made some decisive progress. He made available to scientific research a toolbox, a reservoir of inventiveness to dip into to let us think and organise complicated sets of visible phenomena derived from the experiences of a simpler invisibility. Locating the presence of process morphologies becomes a source of description and understanding, prior to any transformative grasp of the field in question.

To do so, it is important to learn how to discriminate, distinguish and above all separate an underlying form. Was Thom the gatekeeper of the mechanisms through which beings effect individuation? An acte of course separates, according to the *Physics* of Aristotle, whom Thom admired, "*è entéléchéia chôrizei*": division is a morphogenic principle. However, does the Aristotelian notion of separation provide us with the best criterion when it comes to thinking through the individuation of a form from its substrate? A form can be divided from its background: the event contained in any discontinuity leads to a cutting-off from the edge of any form that has thus been individuated. Thom turned to Aristotle when he described an act as "power's edge". ([24], p. 176). But, on the edges, there remains a trace of activity: the regulation constituted by the conflict within the permanent state so as to maintain an individuated entity. From this point, a being or individuated entity is no longer a sufficient departure point when it comes to understanding the process of the genesis of a form. This can also be learned from Simondon: an individuated being rather results from an active process of individuation based on an initial potentiality of being. Before the pre-individual, individuation operates through a dephasing [19].

It seems possible here to detect the working of a morphology of process, in Thom's sense of the term, in other words, one presiding over the internal transformation of one form into another one. It means being able to grasp the notion of becoming in its full meaning of a genesis and to legitimately *bring back time into a form*, which has in the meantime turned back into the form of a dynamic process. This is, at least, the hypothesis which we shall attempt to develop.

2.2 Thom and Simondon Between "Morphology and Individuation"

Morphologie et individuation [26] is the title of a work written by this topologist in homage to the thinker behind individuation and invention. Right from the start, the tension between two works, derived from opposition between a substantialist thesis and an anti-substantialist one, as well as the rejection of hylomorphism, casts deep doubt over any crossed reading. Thom remarked:

> At the beginning, [Simondon's work, *L'individu et sa genèse physio-biologique* (IGPB) or *The Genesis of Individual*], the author seems to have it at heart to adopt an anti-ontologising attitude. An enemy of substance, he invokes the primacy of the relation over the terms of the relationship, which was the central dogma of Charles Renouvier's philosophy! Personally, I have never felt very close to this 'relationist' philosophy: when you have a bridge linking the two banks of a river, the deck is of course the most useful part of the structure, but this bridge also has two abutments which it unites, not to mention the banks and their hinterlands ([26], p. 102)

2.3 Simondon as Read by Thom

However, there can still be brought out from Thom's words several indications that might corroborate the legitimacy of a possible rapprochement between two conceptions anchored in dynamical systems, one presenting an *operation* by which an individuated form is linked to its genesis and the other making a form out of any dynamic genesis.

(i) Firstly, importance is given in these two reasonings to the notion of a potential, be it directly physical (for example, gravitational), or else of a different kind (religious, for example). We shall examine later the notion of a dynamical system for Thom. As for Simondon, he explained the primitive universe by using a comparison with the value of a potential, which is neither just illustrative, in our opinion, nor metaphorical, but which signifies the meaning of the deep-down identity of the morphology of a behaviour that can be generalised in two dynamical systems of a different kind: the shared idea is that of a "difference of functional potential, in an analogy with a difference of gravitational potential, relative to a mass situated at an altitude, whose controlled fall can engender work" ([27], p. 670). This work can be understood mechanically (as action) but also religiously (a symbolic action). This comes down to Thom's notion of transversality:

> The magical universe is structured according to the most primitive and pregnant of organisations: that of the reticulation of the world in privileged places and moments. A privileged place, a place that has a power, is one that draws into it all the strength and efficacy of the domain that it describes, it sums up and contains the force of a compact mass of reality; it sums it up and governs it like a raised site governs and dominates a lowland ([20], p. 164, cité dans Thom [26], p. 101)

(ii) Secondly, in this shift from the religious to the technical as theorised by Simondon, the same set-up appears in Thom:

> "Technicality is opposed to the religious through the *duplication* of the primitive magical centre. As for me, I would see here a kind of 'deployment' of a singularity [...]; I think that Simondon had an intuition about this type of singularity... The first pages of the introduction of IGPB are devoted to a description of this type of behaviour, which he called 'transductive', or 'transduction', which is the generating transformation of this 'spreading of the being' (IPBG, p. 16-18) where the individual is formed ([26], p. 102)

Thom confirms the identity of behaviour between his notion of the "deployment" of a singularity, from which a process will be initiated, and that of a transduction according to Simondon. Each to his own terms, but in the end has not a similar behaviour been indicated, that of a morphological development? Both of them allow for a view of an internal process of transformation allowing any form to access a different level of evolving stability via a treatment *in several phases* of an energetic potential, described as being metastable by Simondon. This notion of phases clearly meets up with that of a process.

Simondon was clear about this: "the pre-individual being is a being without phases, the being after individuation is a phased being" ([20], p. 273, quoted in Thom [26], p. 111). It is for this reason that we should not see an individual as a "a static substance, the relationship should be taken in a temporal sense, between a *pre-individual* being seen as homogenous, and the ultimate state of a *polyphased being*" ([26], p. 102).

(iii) One more indication: Thom suggests considering the existence of a gradual overturning of the theoretical posture adopted in IPGB. According to him, this is one of the difficulties in interpreting this work. "It must not be forgotten that one of the great difficulties of the central thesis of this exposé is that the book (*IGPB*) features a change of orientation between the beginning and the end, between the introduction and the conclusion. Simondon was unable to maintain fully his anti-substantialist thesis." (*Ibid*, p. 104). To the explicitly "mobilising [character] of the Simondonian principle of individuation" (*Ibid,* p. 111) Thom brings back to mind this point and the working principles in action on the limit, in Aristotle's sense, before concluding:

> It is remarkable that our author had discovered, through his study of the allotropic crystalline forms of a substance, the need for a frontier limiting the entity. It is only on pages 112–113 that we find this admission: a crystal is *limited by its edge* of active

accretion in an oversaturated solution. A situation, it would seem, which is quite different from that of the marble statue in Aristotle's metaphysics, a statue whose individuality is defined by the edge of the block of marble cut by the sculptor... (but, for Simondon, this is an insubstantial and immobile edge, it has a 'relational' nature, because something is happening there). Finally, the overall schema of individuation which he considers is effectively hylomorphic, because it involves an interaction between amorphous matter and a 'form' that organises this matter (*Ibid*, p. 107)

2.4 Thomian Forms as Seen in Simondon's Terms

Simondon's notions should be now clarified, as seen in hindsight, so as to develop what can be viewed as a striking closeness between certain properties of Thomian forms and certain properties of individuation. In this respect, which can quite clearly be seen as still programmatic, we shall adopt the particularly enlightening vocabulary used in this philosopher's thought. The aim is to put forward routes for reflexion, rather than setting down a rigorous demonstration. The main point is didactic, so as to enter gradually into the specificity of Thomian forms: as the figures of a process.

(i) Concerning the notion of potential: we shall find again in Thomian morphologies the central notion of potential derived from their mathematical discovery based on a study of the singularities of potential functions. In other words, Thomian figures apply to simple dynamical systems, defined from a gradient of the potential, that indicates the presence of a couple of opposites (high/low, hot/cold, etc.) This notion of the potential lies at the heart of a renewal of Catastrophe Theory in Jacques Viret's research [28]. But, with Simondon, the metastable nature of the pre-individual consists in the presence of such a potentiality.

Individuation presupposes the existence 'of a metastable system rich in potential' (individuation in the light of the notions of form and information (*ILFI*, 27); 'metastable' meaning that it is at the limit between stability and instability), where there exists 'a tension between two disparate realities' (*ILFI*, 31), a potential energy, that is to say a *difference* of potential inside the system [...]; to express this more figuratively, the close and self-maintaining liaison of an overflowing richness and an internal differentiation ([3], p. 47-8)

(ii) Concerning the required differentiation between phases while *entering* an evolving process: we shall present below what should be understood by the figure of a process. But before, let us be attentive to the main characteristics of individuation for Simondon, its polyphased nature, or else the fact that it is impossible to enter an evolving dimension except in the presence of a phase shift.

'*The pre-individual being is a being in which there is no phase*; the being within which an individuation is accomplished is one where a resolution appears through the division of the being into phases, which signifies becoming' (*ILFI*, 67). Becoming is not what happens to the being [...], it is a 'dimension of the being', it is the very operation of individuation in an ongoing process through the 'appearance of phases in the being

> which are phases of the being.' […] In individuation (ontogenesis), the pre-individual being becomes dephased and individuated, but never entirely ([3], p. 48)

This operation of an ongoing individuation restores an internal dimension of becoming which is characteristic of the activity of a process of transformation. As we will see, this dimension can also be seen as a form in Catastrophe Theory but of a rather particular sort: it has lost its edges! It no longer has any edges, strictly speaking, because it attempts to reproduce the heart of a process: fixing, so to speak, what is instable in time (a process) but stable in an evolving dimension. Catastrophe Theory is made up of stable forms, but also and above all of figures that describe the unity of all the phases that make up an evolving process by punctuating the passage of one stable form towards another stable form. The unity of this new type of figure has no meaning as a stabilised identity, but rather, in a sense close to that of Simondon, a rather particular unity since it is figured in an evolving dimension, or what it designates is a transductive unity. It is also thanks to this sort of conceptuality that he treats the question of being in the dynamic sense of an individuation. Furthermore, it specifies the notion, already evoked by Thom, of the spreading of a being, which it should not be forgotten is then associated with his own notion of deployment:

> The conception of being on which this study is based is as follows: a being does not possess a unity of identity, which is that of a stable being in which no transformation is possible: a being possesses a *transductive unity*; in other words, it can become dephased in terms of itself, overflow on one or other side of its *centre*. What we take for a *relationship or duality of principles* is in fact a spreading of the being, which is more than just one unity and more than identity; becoming is a dimension of the being, not what occurs to it as a succession which is to be undergone […]. Individuation should be grasped as the becoming of a being […] (*ILFI*, 31) ([3], p. 32)

If we believe Thom's opinion, quoted above that "Simondon had an intuition about this type of singularity… The first pages of the introduction of IGPB are devoted to a description of this type of behaviour, which he called 'transductive'" ([27], p. 102) then the notion of *centre* evoked by Simondon clearly seems identical in its proprieties to that of an "organising centre" or "germ" of a function for Thom, in other words the critical point from which a process is initiated. It is no necessary to make a more technique excursion into Catastrophe Theory to understand this.

Put simply, Catastrophe Theory is a catalogue of forms whose manifestation occurs in "spaces" with certain dimensions. These dimensions express parameters which are deployed in directions in space, independent one from the other in their signification. In other words, there exists distinct parameters allowing for the deployment of the entire signification contained in the forms. They are not all present in all the figures. Each figure can thus be understood as a partial vision of a broader process that contains them all and of which they express a limited part. They provide successive sections of this process which cannot be represented in its entirety because its dimension is too great: but it is possible to work on certain aspects in local, specific dimensions each time. It is by removing parameters that have remained concealed on some simpler projections of these figures that we can progressively reproduce the signification of the different dimensions of a *morphology of process*. Removing a still-concealed parameter is performed by an operation termed *embedding*, which will be presented subsequently. Each figure can be considered as an organisational centre of a particular change. There are organisational centres containing a very large amount of information (called *umbilics*) and others with a more limited explanatory potential.

It is quite clear that the notion of centre as evoked by Simondon, through which being can be treated, is far subtler than that of Thom which concerns only potential functions. The same goes for the notion of phases. However, we shall show subsequently a morphology of process described in Catastrophe Theory which consists locally in a processual unity of its constituent phases, deployed around an organisational centre that Thom called a generative logos [25]. It is here non-trivial to point out that one of the etymological aspects of *logos*, taken into consideration both by G. Vico and M. Heidegger, lies in the verb *legein*: in his lecture on Aristotle's metaphysics in 1931, the latter noted that "legein means gather (*lesen*) in other words collect, assemble, put one thing in relation with another, consider one with another, place it in a relationship" ([6], p. 15). The notion of relation, of placing in a relationship and proportion returns in the Latin *ratio*.[2]

Apart from the semantic link of *logos* with the general notion of formatting as a placing in relationship of the elements of a whole, the organisational centre is a *logos* in that it generates the different phases of a process, each one placed in relationship with the others: on the one hand, it brings them together at a point, but, on the other and conversely, the phases allow each one to deploy itself in directions all of which are *constituent* of the process in its entirety. Its sole reality is to figure the differentiated integration of the phases in another type of unity. The process expresses their unity, taken in a different dimension: this unity is not played out in the local dimensions in which they can be observed successively and separately, each with their proprieties and in the topological effects that characterise them. To come into being, this unity is played out in a new dimension in which are shown their effects combined in terms of topological retractions, or topological plasticity. These effects are due to the activity of their differentiation, their active differentiation, as in phases deployed in locally distinct directions: what is more, this new dimension is meaningful only as an indication of what results from the deployment of differentiated phases. A phenomenal differentiation occurs through the active work of these phases. They are phenomenologically graspable.[3]

The result is the efficiency of a work of becoming understood as an activity of internal transformation. It cannot be produced without the totality or totalisation of the phases that constitute it. At the same time, each one expresses at a local level the active work of this overall transformation. Each time, this work is one of its constituent aspects. *The understanding of the dynamic operation of this transformation becomes*

[2] Our thanks to Bruno Pinchard for pointing this out to us.

[3] Just as analytical psychology works on the phenomenal content of myths with the help of differentiated energetic polarisations that enter into a relationship of confrontation, fusion or else conjunction, doubtlessly the same should be seen in such developments of the morphologies of processes of a psychic sort [21, 28–30].

the new departure point in which the individuated being is embedded as a series of stable steps participating in a morphology of process. It is at work in each of these individuated realities but is not enclosed there.

Simondon himself was no stranger to these ideas, he was even one of their main proponents, when he noted the co-belonging of a pre-individual and an individuated reality bound in a relationship of *genesis*. As J-Y Chateau explains:

> But, if there is apparently a real difference between [pre-individual and individuated] realities, which are irreducible one to the other, each of them still seems to be irreducibly and wholly the being of the being: 'The being as a being is given entirely in each of its phases' (*ILFI*, 317). How is this possible? It is precisely because this very difference, which is the being of a being, means that the being as a being belongs entirely one to the other not *despite of* but *in* their irreducible difference, what makes them be what each of them is, the same and different: what makes this possible is the notion of *phase*. The pre-individual being and the individuated being are phases of the being. A phase is not 'one temporal moment replaced by another" (from *Mode d'existence des objets techniques [Mode of Existence of Technical Objects]* (*MEOT*), 159), it is only itself in relationship with another phase (or several ones) through the difference of phases, or dephasing. [...] A being as being is a dephasing, a difference of phases, each of the phases being in principle different from the others, and, at the same time, not in solidarity with the others as different parts can be, but being what it is only through its difference with others. We can then see in what way it can definitely be said that a being as a being (dephasing, the differentiation of phases) is given entirely to each phase.
>
> We can also see in this way that it is necessary to understand that this difference in the being, which is the very being of the being, between the pre-individual and individuated being, is real: the difference, which is that of dephasing, is an active difference, a difference that differentiates, the differential through which there is this differentiation [...], which is individuation. It corresponds precisely to the operation of individuation as a process of shifting from the pre-individual to the individuated ([3], p. 34-5)

If there is no difference of phases, there is no possibility to envision this new unity of a transductive type: "the being as a being belongs entirely one to the other not *despite of* but *in* their irreducible difference": The active differentiation of the deployed phases clearly provides a new dimension, which for Thom is a new form, precisely a *morphology of process*, whose meaning remains irreducible to local forms. They are at once a differentiated part of its effectiveness and an enlightenment of its effect in a specific dimension, each one being a particular facet that participates in its constitution: a dynamic formation deployed in an evolving dimension is constituted as an active totality and, inversely, as a totality produced by putting into relationship the morphological work of local forms.

2.5 Towards the Form of a "Morphology of Process"

Let us now explore the way in which Catastrophe Theory allows us to envisage this figure of a process that seems to cross over the notion of an internal development, of the transformation of individuated entities, presiding over their evolution at other levels of scale, based on previous pre-individual levels.

The passage from one stable state to another is typical of what Thom calls a catastrophe: he saw in this mechanism a survival manoeuvre of a system pushed to the limit of its possibilities for regulation. In some situations, it is capable of recovering another regimen of stable functioning: a singular propriety of life of which Catastrophe Theory sets out to provide the geometry. If a catastrophe is intrinsically an expression of a transformation, it can be seen locally as an action aimed at extracting a part of the potential energy available to transform it into useful energy with a view to a "finalistically interesting action", to quote Thom. Through repetition, we can imagine easily that this catastrophist gesture, this action of extraction, can itself become a possible medium, at a different level of scales, with new potential utilisable energy, avalaible for an action on a dynamical system of a different nature. It is in this respect that catastrophes are accompanied by a change of scale, the first sign of an evolution: the kinetic energy of a stream on a mountain slope represents a potential action as available hydraulic energy, even though it is *already* a kinetic energy of movement: it clearly comes from an realisation of a primitive potential, of a gravitational kind, supplying an initial polarisation to the form (the energy field that invests the mountain is not present on the surface of the "relief" of a meteor in zero gravity). However, it turns back into a potentiality in terms of a new type of catastrophe, on a different type of dynamical system: this hydraulic energy becomes utilisable as mechanical energy driving the blades of a watermill. The mill in turn provides available energy for an action at another level of scale: grinding corn. Finally, this grinding will in turn also provide a potential of a metabolic kind for nutrition, etc. The common point between all these mechanisms is, at each level, the possibility of domesticating the resulting dynamical system with the help of what Thom and Bruter have called *a transversality*, which is not without unlike Simondon's transduction. Let us explain this system.

2.6 Summary Topological Description of a Dynamical System and Its Canalisation

The figures of Catastrophe Theory are "varieties", in topology, of forms with a certain number of dimensions. Let us imagine (Fig. 2.1) the surface in relief of a mountain: if this mountain variety is submitted to an earthly gravitational field, for example, we obtain the introduction of an *orientation* and the possibility of having a difference of potential gravitational energy between the top and the bottom, thus establishing a *gradient* of altitude on this variety. Now, if it rains on this mountain, the raindrops will create a privileged field of orientation. They will run along the surface and transform this potential energy into kinetic energy, in motion. This is thus the vision in topology of a dynamical system, in other words *the investment of a Velocity Vectors Field on a Variety* [1]. The derivatives of each raindrop can be interpreted as speed vectors which, seen in their globality, form a field in the physical sense of the term. The speed of the flow will be zero in two *singular* forms of this potential function, illustrated here by the surface of a mountain: at the extrema points, that is to say on the highest plateau (with the *maxima*, called *repulsors*) and right at the bottom (with the *minima*, called *attractors*) in the arrival lakes of the trajectory of movement of the flow of water which becomes stabilised there. The attractors, René Thom's essential quest, are zones of stability for the dynamical systems, as opposed to the repulsors, which are instable. Let us imagine that we observed this landscape long enough for the effects of erosion to become manifest, then the highest points (of mountains or hills) will progressively be effaced, while the lowest points (the lakes) will barely change. Only lakes situated at an altitude might change places (and thus of domain of stability) because of modifications in the beds of the streams because of movements in the ground coming from this erosion. In this case, Thom speaks of a *bifurcation* of basins of attraction, and a *catastrophe* when there is a rapid change in the domain of stability. A catastrophe, in the mathematical (topological) sense of the term, can thus be summed up as an abrupt change of attractors in a dynamical system.

The dynamical system may be considered as an energetic flow that provides energy, but this simply remains available energy, *in fine*, when it is already, in physical terms, an energy of movement: let us stay, as a thought exercise, beside a mountain stream for several hours, this will tell us nothing if we do not act on the stream to *canalise* it with a view to extract a part of it in a useful or utilisable direction within a functioning. We could find at this point a clear distinction between two types of individuation, physical and biological, to adopt Simondon's categories. Landslides

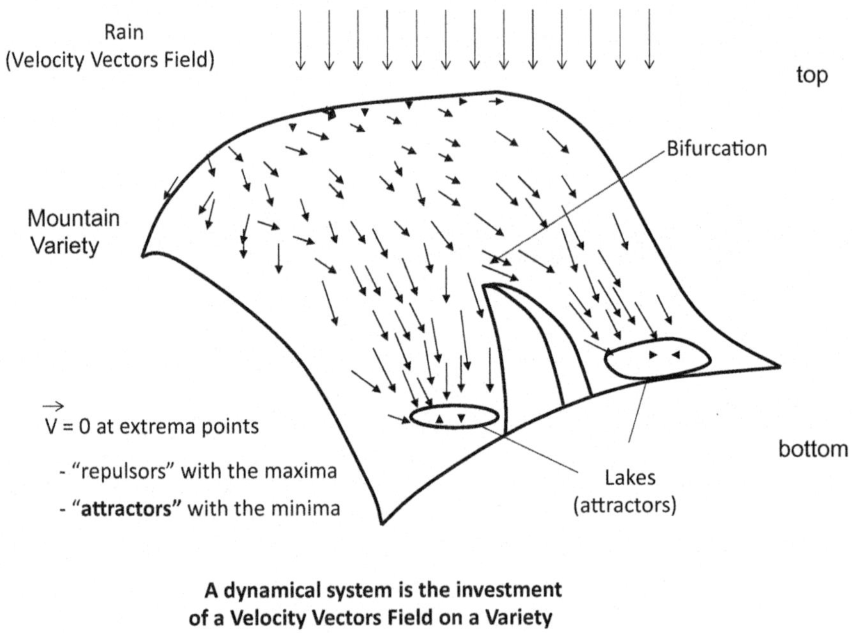

Fig. 2.1 Summary topological definition of a gradient dynamical system

are spontaneous bifurcations, while the organisation of a deviation of the energy flow, so as extract a useful part of it, marks a biological functioning. *This kinetic energy is in fact just the potentiality of an action.* That is to say, this dynamical system means having an energy of movement, of course, but it remains realisable. It will thus only be realised if we *act on* the stream by what Bruter calls the *transversality* and Thom the *canalisation* of a dynamical system (Fig. 2.2). It is only here that this energy, which was previously only available, becomes useful. The varieties considered by Thom are initially polarised, in our example by the presence of a Gravitational Force Field, but more generally be *pregnancies*. Thus, the rain procures a *materialisation* of this pregnancy and supplies, through a flow over the polarised variety, an energy of movement: this becomes the true *functional potential* of the dynamical system.

In more literary terms, the dynamical system appears when the form is crossed over, or affected, invested to sum up, by a certain flow or the excitatory pregnancy of a domain with a privileged orientation– the Velocity Vectors Field– and which is a surface which will start to flow. The morphologies of Catastrophe Theory will be the limited parts of this energetic landscape where a restricted number of attractors enter into action. In an evolving system, "catastrophes" allow for the canalisation of this functional potential through the exercise of a bifurcation of the flow. Let us now pay attention to this notion of "evolution" present in both Thom and Simondon.

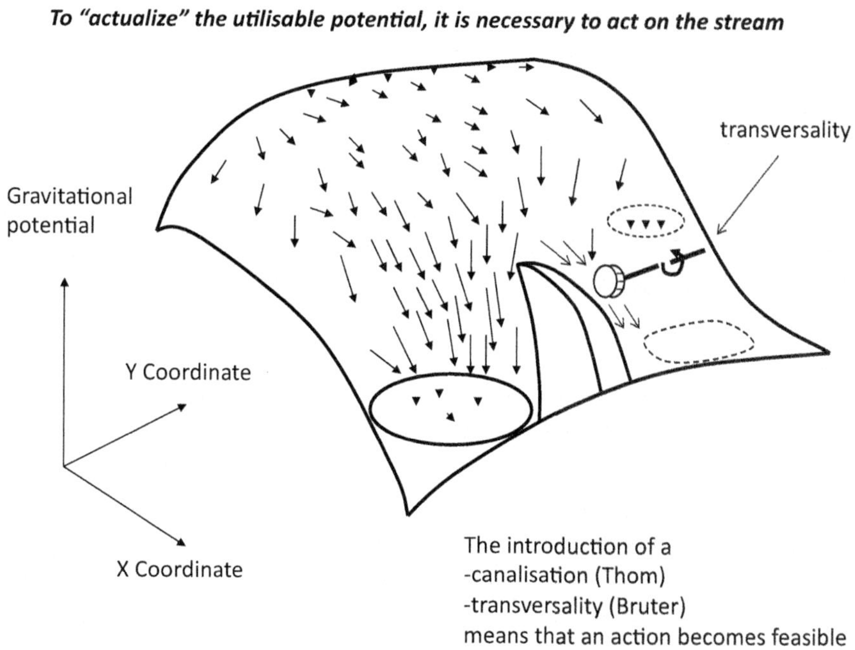

Fig. 2.2 The introduction of a transversality in a dynamical system allows for the extraction of useful energy

2.7 Two Elementary Catastrophes: From the Fold to the Swallowtail

Of course, the approaches are distinct, one from the other, and it is necessary to enter into the figures of Catastrophe Theory to grasp in what way they express even a local functioning, which contains a dimension of internal temporality. From a catalogue of figures, all stable, Thom allows us to consider the passage between these figures, which is stable too.

The functioning begins, as with Simondon, in a homogenous state: which he describes as being metastable, it covers a potentiality from which individuation can be derived. A common point with Catastrophe Theory: the point of departure of the process is found on a figure representing a state which is stable, but only for a time, because it will be caught up by wear, a progressive disorganisation brought in by an entropic-type parameter which will, if it increases, end up engendering the first of Thom's "catastrophes". In fact, this is phenomenologically a discontinuity, from which a dynamic has bifurcated, but as opposed to other Thomian catastrophes, its particularity is that it engenders no useful work, no form of individuation in a sense, given that the dynamic leaves a stable state without finding any other form of stability at a different level of energetic functioning, which would have expressed a transformation produced by a mechanism of catastrophe.

Being stable for a time, a particular domain can be seen as a figure which is a well of potential equipped with a phenomenal discontinuity. We can see the singularity, the critical point, as the extremum of the function, to the left of Fig. 2.3, materialised on the right by the line of folding. If an energetic flow invests this variety, we obtain a dynamical system. It can only follow the naturel flow of the slope. The disappearance of its basin of attraction occurs along the folding line of the leaves.

Let us begin with a membrane or elastic band, pulled and folded, whose lateral sides have been fixed to the extremes of a support. If we stretch this band with one finger,[4] (a band which we suppose to be infinitely stretchable without being torn, as it is continuous) there will be no *qualitative modification*: this form is thus qualitatively *stable* and this is the first of the seven "catastrophist" morphologies in Catastrophe Theory, called "the fold". Let us posit a ball lodged on the upper leaf: it could stay there for a certain time. Leaf A or B represents a domain of stability in the dynamical system being studied, represented here by the movement of a ball. By disturbing slightly its position on this leaf, it could descend a little but will remain on this stably installed surface. If pushed too hard, it will then cross over the line of the folding and fall beyond the limits of its domain of existence, viability and stability. This fold evokes the presence of a single form of functioning. "Let us imagine that this system is a stably functioning boiler, and we increase the heating pressure. A moment will come when the boiler explodes. This is a catastrophe in the usual sense of the term, of course, but once again, not in the mathematical sense, because the boiler, now destroyed, can no longer find another mode of stable functioning" [29].

[4] The original idea of using the fingers to make for a comfortable visualisation is due to Jacques Demongeot.

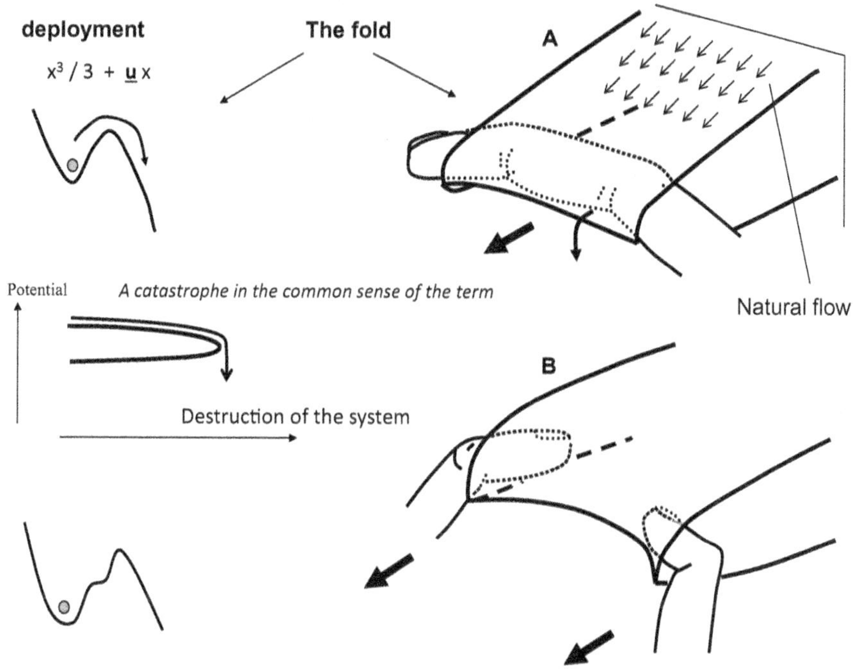

Fig. 2.3 The characteristics of the morphology of the fold

This figure can begin to undergo an internal modification as soon as a polarity is introduced (B). In this respect, a new dimension already contained in the fold, but in a state of virtuality, may be deployed. Let us work on this traction to observe its effects. And pursue the morphological evolution of the fold being stretched into two distinct polarities (Fig. 2.4).

The effect of this stretching is to produce a structuration of this form, which has become destabilised by these tractions (at B on Fig. 2.3 firstly, then at C and at D on Fig. 2.4). If the structuration increases beyond a certain limit, there will then be a breaking of the symmetry, the appearance of a partial coverage zone of one of the leaves over the other, one of the two attracting structural trends thus becoming dominant (E). This zone, via a supplementary stretching, will be able to be extended in the direction of this structuration, but nothing else will happen qualitatively from this topological transformation: this form has thus once again been stabilised. On the other hand, the progressive structuration of this form gives us the possibility to obtain an additional piece of information, which is deployed transversally to its structural organisation: here too, another new dimension arises. This dimension is realisable only at the level of the coverage zone, because if a slight vertical impulsion is given this time, it becomes possible to change alternatively the order of dominance of the attractors (in Fig. 2.4, from E to F, then from F to E, etc.).

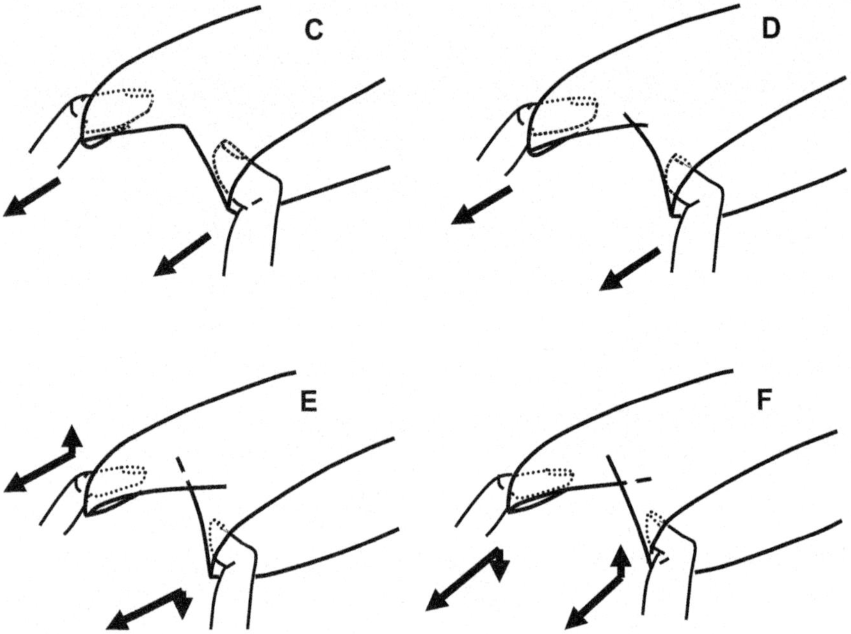

Fig. 2.4 Evolution towards the swallowtail via a progressive structuration of the fold: appearance of a possible zone of transversality

There can then be in this zone a *bifurcation*, or sudden change of dominant attractor in a dynamical system, which is the mathematical definition of a catastrophe. In Fig. 2.5, we resume the parameters organising the deployment space of this new form, called a "swallowtail", according to an initial *structural* variable, a second variable of a *functional* sort, transversally disposed on the structuration, through which regulatory actions on the stability of the form can be made discontinuously. These actions are made by means of an energy exchange transferred from the dominant to the dominated attractor. Finally, a third dimension makes a difference at a *potential* level, developing vertically, being progressively generated and organised from the rear towards the front of the form thanks to the structuration. The functionality can then use this potentiality to carry out a useful action.

Thus, via a section along the structural dimension of the fold, Fig. 2.3, we noted the trajectory of a catastrophe in the common sense of the term, leading to the destruction of the system. If we now effect a section in the direction of the transversality at the coverage zone of the attractors, we will obtain a projection in two dimensions (potentiality and functionality) of the form most often worked on by Thom, contained in this figure and called a "cusp" (Fig. 2.6).

Above, the mathematical view: the dynamical system, via an internal deformation of the function, is required to leave the state of stability where it is: from (a) progressively deformed into (b) then (c).

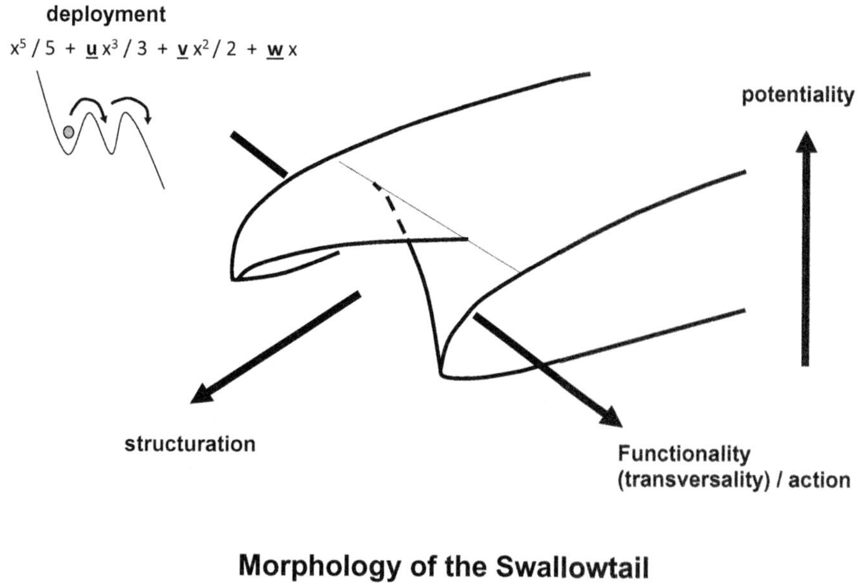

Fig. 2.5 Parameters of the deployment of the swallowtail

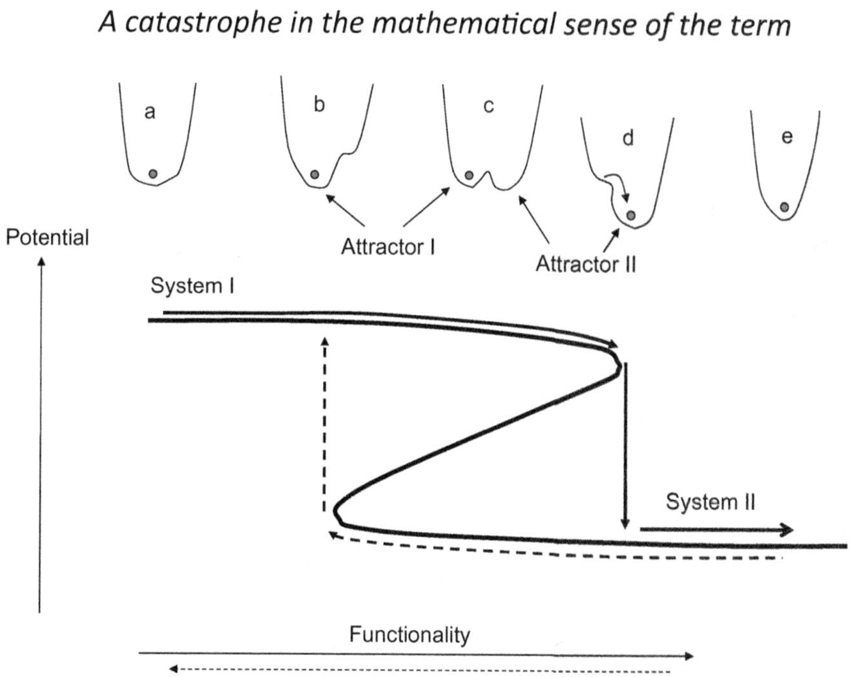

Fig. 2.6 Section of the coverage zone of the attractors: the swallowtail contains the cusp

Below, we can observe the (simplified) section effected in the coverage zone, as in a cube game, the cusp is formed from the fusion of two folds (section of Fig. 2.3). Once at the level of stability of attractor II, the system may locally be in a new state, but through a *drop* in potential. At this moment, a new datum to be taken into account in the biological individuation in terms of physics appears: a rising trajectory (dotted lines) via *hysteresis* reminds us of the thermostat of a heater programmed to come on and off after the crossing of a falling or rising threshold in the gradient of the ambient temperature.

2.8 The Phases of the Process and the Dimension of Internal Transformation

However, in the living world, a return trajectory at the level of the initial potential can only be made through an addition of energy linked to the activation of a second dynamical system to the signification of a *refractory* metabolic flow to the flow of action. The refractory step and its phases occur *in the opposite direction* to the step of action and its phases. In Fig. 2.7, we reproduce a simplified view of an energetic function, long worked on by Jacques Viret.

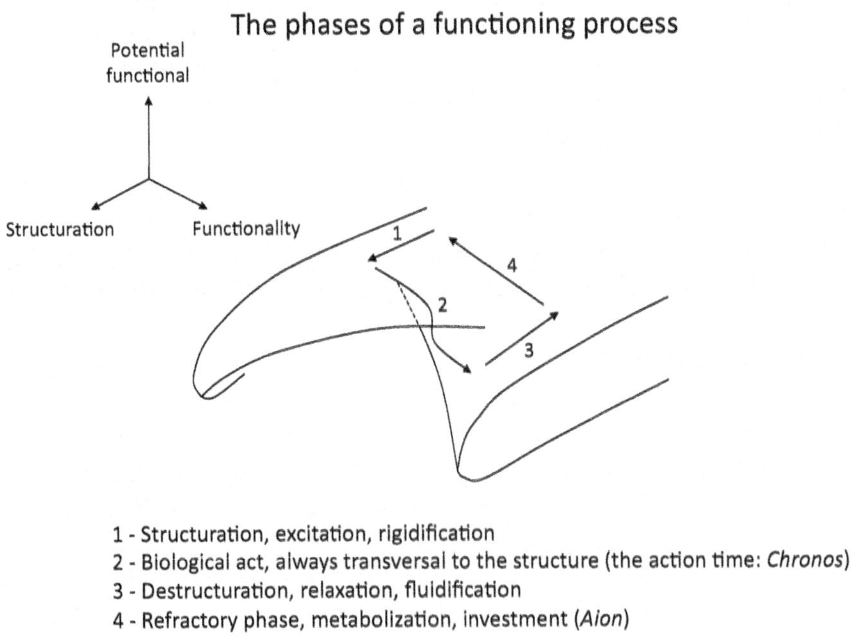

Fig. 2.7 The unrolling of differentiation of the phases of a biological function according to Jacques Viret

To the phase of structuration of the attractors, from the fold towards the swallowtail, the response is a phase of fusion of the attractors by phenomenological de-differentiation, or fluidification of the previously rigidified couple of opposites: this phase allows for a return to undifferentiation (Simondon's pre-individual) via a destructuration of the conditions of activity of the energetic system. At this structural phase which might be called "regressive", to use the language of psychology [8] a functional phase will be initiated in the opposite direction in turn to the phase of action, like a sort of investment inverse to the expenditure: its signification is a "metabolization" that allows for an internal reorganisation of the system, prior to the initialisation of a new process. We might ask if Simondon's pre-individual does not acquire here a crucial role: it will, in a way, be a link between the previous and the subsequent steps, as the zone of the fold is the point of departure and of arrival of an evolving function.

A few remarks at this level of the presentation of Thomian figures:

(i) A being, initially homogenous, undifferentiated by attractors, still has a "capacity for structuration" (fold, in A). There is a case to conceive the existence, alongside a functional potentiality, of a structural capacity: it is necessary to create the possible structure for the realisation of the functional potentiality in action. Simondon's pre-individual, like a reservoir of ever-new differentiations, would be the place where this capacity could be derived. We thus see that the individuated being is not in the figures of Catastrophe Theory the first element in the process that presides over its formation.

(ii) The transversalities are accompanied by a change of level. Without multiplying swallowtail figures, we can understand this phenomenon simply from this very figure. The phases in the process mentioned above reveal alternatively structural and functional dimensions allowing for the use or the metabolization of a functional potential.

(iii) We are thus witnessing a to-and-fro phenomenon, which is successive and alternating, between the fold and the swallowtail. The realisation of this process does not exclude temporary involutions according to a mechanism alternating progression and regression, which is rather reminiscent, as in Goethe, of a *diastolic* and *systolic* pulsation. "*So that a stable object may evolve*, it has to pass, before reaching a new state, through a phase of 'local regression', during which it is partly destructured. […] The object recovers the qualities of its youth, its plastic qualities, which make it capable of taking new routes of evolution" [1]. This is a form of totipotence.

Another aspect of this swallowtail then occurs, which is a figure with four dimensions, not discernible to the sight in three dimensions, while bringing in the action of an evolving parameter. To raise this hidden variable, an embedding has to be made: this means hiding one of the dimensions of the object to observe its behaviour in another one, which is invisible or hidden until then.

The illustration (Fig. 2.8) of the evolution of an explosion by a circle that grows over time into a cone works by just such a procedure: the point is to hide one of the dimensions of the ball (variety in a volume with three dimensions) thus becoming a

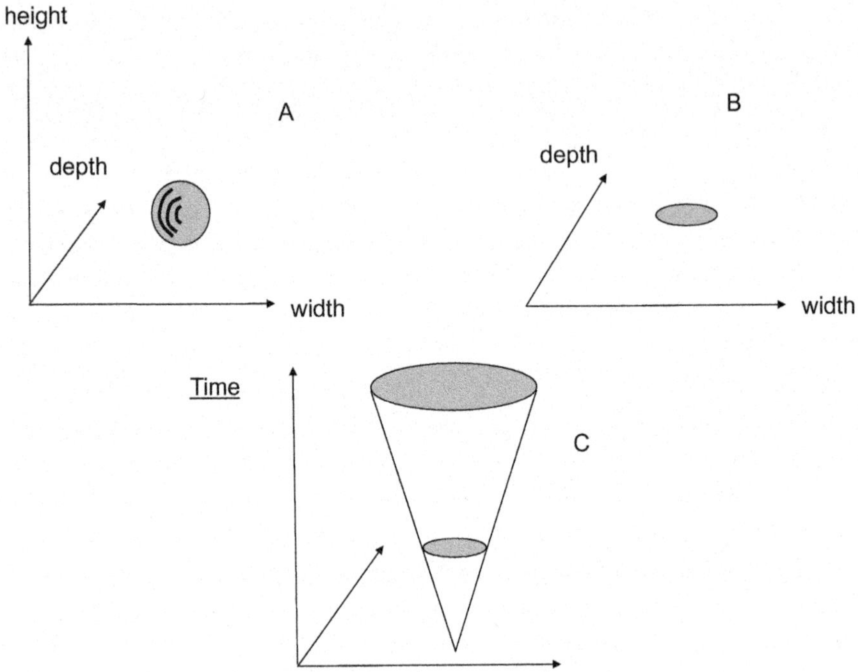

Fig. 2.8 Evolution of the ball via a circle that grows over time as a cone

circle (variety on a plane of two dimensions). The height is suppressed and replaced by time. We thus observe the growth of the circle as a cone that figures the evolution of the initial ball.

We might repeat here the operation for the process that leads from the fold to the swallowtail. We suppress the depth of each figure to keep just the signatures of the catastrophes on the plane (Fig. 2.9).

On the right, the succession of the figures already encountered. In the centre, the discontinuous edge of each of them. When we connect all the points, we obtain on the left-hand figure a new representation of the swallowtail's morphology of process. Thom directly proposed this representation in his manual [22] without passing via the figures on the right, but without extracting all the significations, being set apparently on enlightening the scientific community of the day about the implications of his discovery: the signification of catastrophes in the mathematical sense of the term. The following point should be admitted: the extreme density of his formulation of a "*generative logos*" lies at the root of the signification of this form. This expression condenses like an organisational centre what we are trying to deploy laboriously. It makes it fulgurant in an instant.

The figure on the left has no more edges, strictly speaking, but expresses the heart of a process. It is made up of the unity of differentiated phases on the figures, whose activity creates topological retractions deforming the plasticity of this metabolic

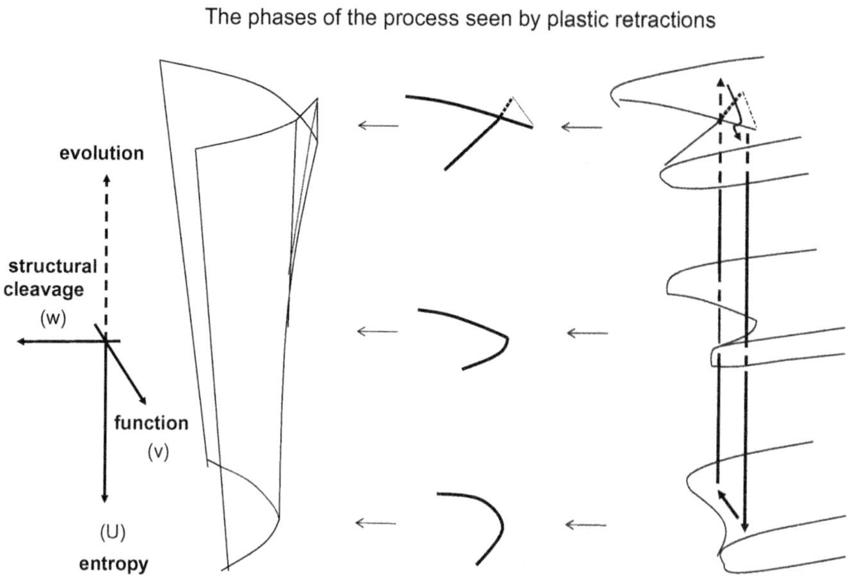

Fig. 2.9 Figure (topology) of the process of evolution by the internal transformation of one figure into another: the swallowtail as a morphology of process

space. This is no longer the usual space which cannot express time by these figures or configurations, but instead the "figure" or the effective reality of a process. It is in this precise sense that the vertically developed dimension has an evolving temporality, unexempt from involution, something which a Galilean duration cannot feature. Here, time is coupled with changes or transformations of a state which modifies its signification. In Simondon's terms, this would be a *transductive* unity, not that of the identity of a stable being, but that of a dynamic genesis. Can the dialogue between Thom and Simondon be picked up again by research exploring the meaning of a figure that Thom never really worked on? A catastrophe, as we have said, means a mutation. But it seems all the more convincing to bring out the proprieties of this process of transformation by studying qualitatively the proprieties of the plastic deformations that it engenders in the interior of the topology itself.

We have not been able, in this exposé, to explore the precise aspects of these deformations, nor even unveil all the local facets of this morphology of process, of which the swallowtail is just one projection. However, it is still revelatory at the level of this morphological notion of the phasing of a being: *the being of the swallowtail is this unity of differentiated phases*, in which "each one is only what it is through its difference from others", this differentiation being active but conjoined in a single form, that of a morphology of process.

Simondon's *mobilism* participates in a facet of the history of thought, arising precisely in a reaction to the *substantialism* supported by Thom. However, the latter opened a fecund route by working on the objectivity of the processes of metamorphosis, from which we might glimpse the nature of embedding, both in its mobilist and substantialist aspects. This line of thought is enlightened by an exchange between Bruno Pinchard and René Thom. The latter concludes concerning the philosopher's attempt: "The issue raised in *La Raison dédoublée* [a work born of B. Pinchard's thesis] is that of change, of transition. The author raises in it the problem of metamorphosis: how to figure a transition, which is in principle discrete, but on a continuous connected medium, from one entity to another?" (Thom 1992, p. 596) B. Pinchard mentions: "And Thom then proposed, to solve this purely topological problem, to resort to 'a (new?) form of the *analogy of being*, which would identify the being of the substance with the being of the transition, the being of the metamorphosis'" ([17], p. 604–605).

Two lines of thought could indicate a programme for work around the concept of *embedding*.

(i) This does not seem to come down just to a mechanism or a topological technique: it seems to feature an efficacy based on an internal transformation, which certain fields such as analytical psychology, for example, take as the heart of their curative practices. Psychic unlocking pays close attention to the alternation of temporary episodes of involution required for any subsequent evolution.

(ii) Such functioning can also be explored in the field of politics, through a treatment of the memory envisaged collectively, in its symbolic dimension. The bonds of symbolisation between *memory* and *history* in the "sites and knots of memory" bring into play the (energetic?) investment of emotions and affects, leaving behind certain nodal collective sites during crucial periods of symbolic reorganisation, while inaugurating a mechanism of reinvesting new sites. This research into such mechanisms is linked to the multiplication of work in the social sciences on the notion of *historicity* which attempts to reintroduce *time* (Clio) in the *structure*, after its intellectual effervescence then in its decline in French intellectual thought during the 1950-1980s [4]. We consider that this historicity marks clearly a dimension of internal transformation present in what Marcel Gauchet has called the *political institution of human-social establishments* [5, 12]. These questions seem to be fields for study for a morphological relevance, in the sciences of humanity and society…

References

1. Bruter, C.P.: Les architectures du feu: Considérations sur les modèles. Flammarion, Paris (1982)
2. Bruter, C.P.: Les formes du continu. In M. Espinoza (dir.), De la science à la philosophie. Hommage à Jean Largeault (pp. 61–78). L'Harmattan, Paris (2001)
3. Château, J.-Y.: Le vocabulaire de Simondon. Ellipses, Paris (2008)
4. Dosse, F.: Histoire du structuralisme. 2 tomes: I Le champ du signe 1945–1966, II Le chant du cygne 1967 à nos jours. La Découverte, Paris (1991, 1992)
5. Gauchet, M.: La condition politique. Gallimard, Paris (2005)
6. Heidegger, M.: Aristote, Métaphysique Θ 1–3. De l'essence et de la réalité de la force. Cours 1931. Gallimard, Paris (1991)

7. Largeault, J.: Principes classiques d'interprétation de la nature. Vrin, Paris (1988)
8. Lavallée, G.: Régrédience, progrédience et hallucinatoire de transfert. Conférence d'introduction à la Psychanalyse de l'adulte du 20 Janvier 2005. [SPP], Paris (2005)
9. Leibniz, G.W.: Le discours de métaphysique. Gallimard, Paris (1686)
10. Morier, C.: Théorie des catastrophes et science politique: Éclairage de la théorie du politique chez Jean Baechler par les morphologies de René Thom: vers un structuralisme dynamique. Editions Universitaires Européennes, Saarbrücken (2011)
11. Morier, C.: Taking a geometric look at the socio-political functioning schemes of the living. Catastrophe theory and theoretical sociology. Acta Biotheor. **61**(3), 353–365 (2013)
12. Morier, C.: Les morphologies du politique. Approche comparée des œuvres de René Thom et Marcel Gauchet. LGDJ, Paris (2018)
13. Petitot, J.: Morphogénèse du Sens I: Pour un schématisme de la structure. Presses Universitaires de France, Paris (1985)
14. Petitot, J.: Physique du Sens, De la théorie des singularités aux structures sémio-narratives. CNRS Editions, Paris (1992)
15. Petitot, J.: L'idéalisme du rationalisme physique au réalisme de la philosophie de la nature. In: Espinoza, M. (dir.) De la science à la philosophie. Hommage à Jean Largeault, pp. 47–60. L'Harmattan, Paris (2001)
16. Petitot, J.: Morphologie et esthétique, La forme et le Sens chez Goethe, Lessing, Lévi-Strauss, Kant, Valéry, Husserl, Eco, Proust, Stendhal. Maisonneuve et Larose, Paris (2004)
17. Pinchard, B.: La raison dédoublée: La fabricca della mente. Aubier, Paris (1992)
18. Pomian, K. (dir.): La querelle du déterminisme. Gallimard, Paris (1990)
19. Simondon, G.: L'individu et sa genèse physio-biologique. Presses Universitaires de France, Paris (1964)
20. Simondon, G.: Du Mode d'existence des objets techniques. Aubier, Paris (1969)
21. Solié, P.: Le sacrifice: fondateur de civilisation et d'individuation. Albin Michel, Paris (1988)
22. Thom, R.: Stabilité structurelle et morphogenèse. Intereditions, Paris (1972)
23. Thom, R.: Modèles mathématiques de la morphogenèse. C. Bourgeois, Paris (1974)
24. Thom, R.: Esquisse d'une sémiophysique, Physique aristotélicienne et Théorie des Catastrophes. Intereditions, Paris (1989)
25. Thom, R.: Apologie du Logos. Hachette, Paris (1990)
26. Thom, R.: Morphologie et Individuation. In G. Hottois (et. al) Gilbert Simondon. Une pensée de l'individuation et de la technique. Albin Michel, Paris (1994)
27. Viret, J.: La fonction physiologique. In: M. Porte (dir.) Passion des formes, Dynamique qualitative sémiophysique et intelligibilité, pp. 661–679. ENS Edition, Fontenay-St-Cloud (1994)
28. Viret, J.: Thom et Jung, un dialogue imaginaire. Psyché et Théorie des catastrophes. Pantin, les éditions Baghera (2018)
29. Viret, J.: La notion de l'archétype chez Jung et dans la théorie mathématique de René Thom. In: Baum, C.Y., Decharneux, B., Szafran, W. (dir.) Jung et les sciences, pp. 97–117. E.M.E, Fernelmont (2009)
30. Viret, J.: Topological approach of Jungian psychology. Acta Biotheor. Biotheor. **58**(2–3), 233–245 (2010)
31. Zeeman, E.C.: Catastrophe Theory, Selected Papers, 1972–1977. Addison-Weysley, London (1977)

Chapter 3
Chorematics. The Dynamics of Lived Space

Per Aage Brandt

The Greek term *chora*, χωρα, designates the location, the place, the space where one is, or one leaves, etc., in short, space as it is experienced.[1] Space experienced and semiotized by this fact: named, interpreted, it is a dynamical entity, to which we lend strengths and forms.[2] We can call this dynamic form taken by an elementary *chora* a *choreme* and thus propose to develop a semio-spatial analysis, a *chorematics*.[3] Everything of course depends on what we will find to retain in the fields, elementary or more complex, of spatial and spatio-temporal experience, by examining it from this formal, but phenomenological point of view. A place is a closed space, surrounded by a more or less clear border, created by separating them into an interior and an exterior, an inside and an outside. The prototype is probably the dwelling, the protected place where we live, behind walls that stop the dysphoric flows that approach from the outside (rain, storm, malevolence, predation) and that retain the euphoric flows and the objects that exist within them, as well as the subjects within them. In this sense,

[1] *May this article celebrate the memory of our friend and mentor.*

[2] This article was translated by Jacques Demongeot and reviewed by Wolfgang Wildgen.

[3] Our first elaboration of a *chorematics*, or logic of places, dates from the book Sandheden, *sætningen og døden* [5]. At the time, this primordial version was going to make it possible to integrate the dynamic semiotics of René Thom [15] and the cognitive semantics of Eve Sweetser and Leonard Talmy [13, 14], that is to say the catastrophic model and the force-barrier model, in a research paradigm that was then presented under the term of *dynamic and cognitive semiotics*. The Platonic notion of *chora* has also given rise to geographical *chorematics* [6] and to architectural [7] and psycho-semiotic [10] reflections. The latter are at the origin of our idea of a cognitive *chorematics*, a fundamental schematism of the place, underlying the notion of "container" as metaphor or image-schema of Lakoff and Johnson [11].

Per Aage Brandt (1944–2021).

P. A. Brandt (✉)
Case Western Reserve University, Ohio, United States of America
e-mail: p900292@ulusofona.pt

© The Author(s), under exclusive license to Springer Nature Switzerland AG 2024
I. Marcos and C. Morier (eds.), *The Relevance of René Thom*, Lecture Notes in Morphogenesis, https://doi.org/10.1007/978-3-031-54982-3_3

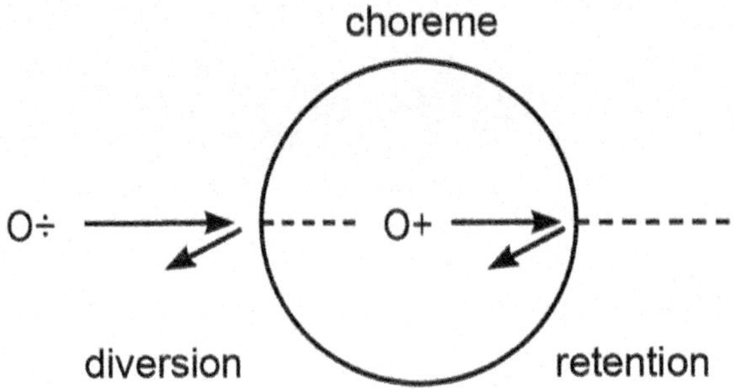

Fig. 3.1 The two senses of protection. Diversion and retention

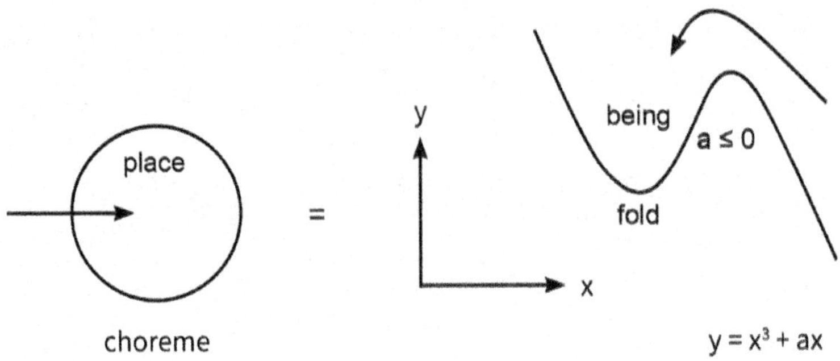

Fig. 3.2 Correspondence *choreme*—fold

the chorem protects. It diverts, and it retains, by its edge functioning as a *repeller* (Fig. 3.1).

The importance of the choremic border as a double barrier, a repulsive force, or even a deterrent, is obvious. The barrier is a repeller, while the interior is an attractor; the 3D *choreme* can therefore have the dynamic shape of a crater or a bubble, a balloon. In terms of catastrophic topology, it simply corresponds to a *fold* (potential $V = x^3$).[4] This elementary catastrophe represents the *being* for Thom. If the fold is smoothed out, *being* is no longer (it is nothingness). The fold has only one variation parameter, the control variable a in the equation: $V = x^3 + ax$. The minimum, i.e., the attractor, exists if a is negative. In this case, the barrier, the repeller, also exists (Fig. 3.2).[5]

---

[4] Figuratively, the *chorem* is a fold rotating on its minimum, so to speak, which makes it compatible with the three-dimensional space of phenophysics where it can thus be a "place".

[5] A minimum is an attractor for flows that seek structural stability requiring to spent a minimum of energy; it is this convention that also makes the maximum a threshold and a barrier, a repeller [15, 16]. The barrier is indifferent to the difference between what comes from inside and what comes from outside of this *choreme*/fold that it defines by rolling over itself, as a guardian of the closed. *Protection works both ways.*

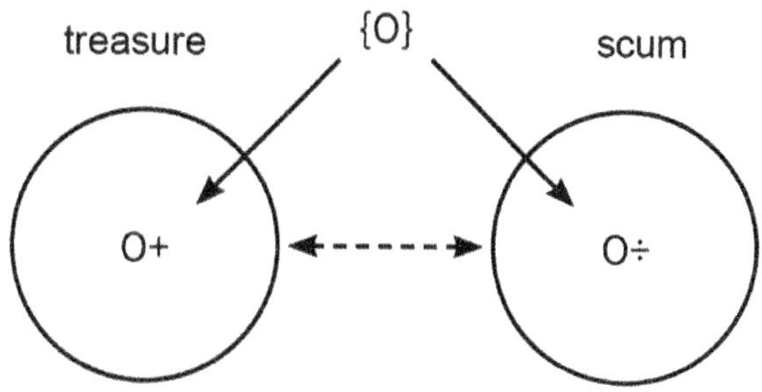

Fig. 3.3 A simple polar *chorematics*

To value is necessarily to internalize, to retain; in this sense, it is to hoard the O+, which presupposes a place protected for the *treasure*, its *choreme*. Devaluing, on the other hand, is an act that reverses this spatial logic: one invents a negative, external place, and one locates there what is depreciated (O÷); a landfill, a scrap, a prison, examples abound. Things thus have their place, from where they can nevertheless be moved, which gives rise to stories (Fig. 3.3).

This is how we semantize physical space, and cover the world with geopolitical and ideological specifications: *the same and the other* are the entities separated by the walls of the places, all in a neutral space, open. The polar arrangement seems above all to serve ideological and narrative purposes: opposing two contrary *choremes*, therefore two interiors, is already more dramatic than simply opposing a valued interior and a neutral exterior.[6]

If two folds meet in space in this way, they can approach each other and combine topologically to form a small local universe with two attractors. The two wells then become conflicting forces, and this formation becomes agonistic. This is what happens when two *choremes* touch and overlap. Topologically, we then have what corresponds to the cusp catastrophe: Unfolding $V = x^4 + ax^2 + bx$, where a and b are the control variables (Fig. 3.4).

This conflict[7] can give rise to an exchange or a struggle, in the sense that the two attractors share or dispute the energy flows that cross the barriers. The choremic intersection can even turn into a stable attractor, and we will then have a local topology with three attractors ($V = x^6$), the butterfly catastrophe.[8]

[6] The semiotic square by Greimas ad Courtés [8] corresponds to this simple *chorematics*.

[7] The circles S1–S2 of Fig. 3.5 which intersect are *choremes*, but could be read as logical sets and Euler circles. However, as my colleague Wolfgang Wildgen very pertinently points out, in a commentary on this text for which I thank him, logical relations have no spatiality; the interest of this *chorematics* is, on the other hand, to give back to the natural logic of our cognition its spatiality, and thereby, its dynamic dimension. The formalizations of the logical type are *chorematics* which are unaware of each other.

[8] We recently had the opportunity to use the butterfly model to account for the dynamics of enunciation (Actes Sémiotiques, No. 123, 2020, Accessible on the internet). We explain in the mentioned article the unfolding of this catastrophe.

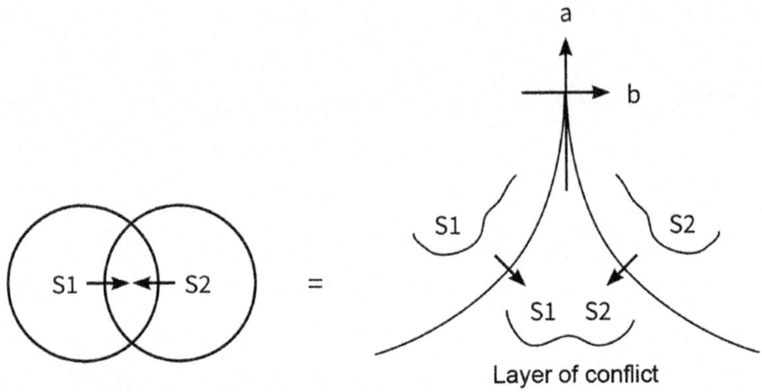

Fig. 3.4 *Choremic* conflict, formation of a cusp

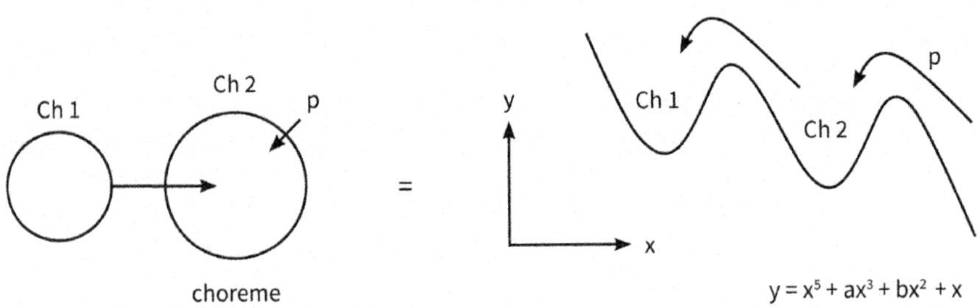

Fig. 3.5 Choremic embedding. Fold in the fold, swallowtail[9]

In the lived space, we encounter this choremic conflict in sports games of ball and ball, or in the game of chess. It is also present in the spatial arrangement of political parliaments (left—center—right).

Agonistic games always spatialize explicitly, and they manifest another very important syntactic feature: framing. Consider a strongly demarcated playing field. This framework constitutes in itself an encompassing *choreme*; it is therefore necessary to understand how one *choreme* can encompass another. In this case, there will be a fold within the fold, so that you cannot access the first without going through the second. The potential $V = x^3$ of the fold then becomes a potential $V = x^5$, and the fold becomes a swallowtail. Embedding semiotically represents the conditional figure: a space and what it contains or does not contain determines or conditions another space it embeds. A field of possibilities thus determines what becomes a subfield of possibilities.

[9] The swallowtail control topology is discussed in a narrative interpretation, in our *Modal Framework of Meaning* [4]. It is the deformation path crossing this deployment topology that decides the assignable meaning, because this path characterizes the time of the change in the spatial condition of the entities traversing it. These entities are represented by the "p" of the graph of Fig. 3.5.

We can now consider *choremes* that enter into other *choremes*, of higher order, or that come out of them. The dynamics of those who move will then be affected by those who receive them or let them go. If a *choreme* is a place impregnated by a certain character, a quality determining what it contains, the mobile *choreme* that enters it will in turn be impregnated by this character, this quality. This is what gives rise (as we say so well) to propositional semantics: the Subject *choreme* enters the Predicate *choreme*, S 'becomes' therefore P and 'will remain' P until leaving it—then S will 'be' no more P. The predicative negation ¬P(S)—S is not P—expresses the exteriority between S and P. If S is P, it will be determined by the meaning of P, that is, say by the condition that P imposes on what has entered it. If one thinks in this way, a small spatial–temporal and dynamic logic settles in the lived world. The conditionality follows directly: If P contains the quality rain, S will come out wet. If P is a temple, S may emerge blessed and touched by the magic of the deity. If S is in P, and P contains the substance Q, then S exposes itself to the effect of Q. It can obviously still protect itself against Q if it enters an anti-Q, R, before or afterwards, and that thus it is encapsulated in R while it crosses P. We then have a constellation {P(R(S))}, easy to exemplify: epidemic, mask, subject.

Dynamic schematisms know no scale: whether it is a question of a minimal daily phenomenon (such as the logic of an umbrella) or of a concept of great extension such as the management of a vast territory, the thought uses the same forms.

3.1 Social Chorematics

Thus, we can consider that a society is composed of several collective components which condition each other. Schematically, we have at least a model with three instances, or layers, or again, therefore, *choremes*[10]: a fundamental layer of production of goods and food, a layer of protection of the community, and between the two, an institutional layer—which can be covered with variable terminologies. There is thus an organic component which provides food and which is essential to the very existence of a population territorialized by agriculture (in prehistory, around –8000); let's say that these are the beings of the earth. They are fragile, and conflicts between tribes are frequent and deadly. However, since the Bronze Age (–3000), this organic *choreme* enters a protective bubble, a symbolic *choreme* composed of warriors and priests who respect the spirituality of the protective violence offered. They are beings of the metal.[11] From now on, the beings of the earth are therefore protected by the beings of metal. The new problem that immediately arises is that these protectors also exploit and terrorize the population; it is therefore necessary to protect oneself from it again, which is done by inserting a sphere supposed to regularize the relations

[10] We use here the ecological model developed in "The Meaning and Madness of Money" [2].

[11] Namely, the metal of weapons, insignia of religion and money (metal currency): barracks, temples and banks are often mixed or merged.

between the metal and the earth, between the sacred violence and the organic population. This interposed sphere corresponds to the emergence of legislation, city-towns, a "public" life, based on the extraction of a third ecological category after food and metal, stone. The beings of the stone, today of the urban concrete, intercalate and invent civic regimes, prevent the sacred warriors from ransacking the population, now partly become "citizens". It is the origin of public and political life; an inevitable polarization will now socially oppose a warrior 'right' that could weaken the walls of stone to allow the symbolic to reach the organic, as seen in authoritarian regimes and populisms today, and a 'left' which seeks to support the stone instead, and thus support the protection of the population against the threat of predatory 'protection'. In times of war, the stone is silent, and the symbolic is supported everywhere; in times of peace, on the other hand, the political discourse of stone flourishes. This schematic history structuring the social body can be summed up in two integration operations, those shown in Fig. 3.6.

The choremes move easily, and the organic can for example partly overtake the political, thus finding itself directly under the symbolic; thus, the legalist regime can coexist with a fascist populist anti-regime. This model probably corresponds to an intuition widely shared in the world community which, in the current situation, has the opportunity to compare and evaluate the state of nations and regions of the planet.

The role of the political sphere is thus to insert itself between symbolic violence and the organic base, and to constitute on the one hand a space of communication, speech and exchanges inscribing the organic in a civic environment allowing the regulation behaviors and deliberation on common problems, and, on the other hand, a bulwark against the agonistic arbitrariness of warriors and religions, in short a democratic space in the deep and elementary, even rudimentary, structural sense. The new problem it introduces is obviously that professionals in the political sphere have

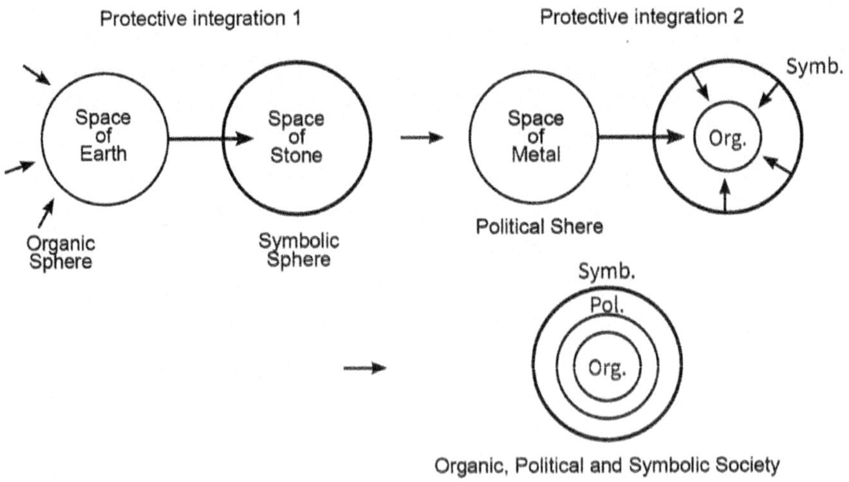

Fig. 3.6 Social dynamics of Earth, metal and stone

to navigate between their internal (organic) 'citizens' and their external (symbolic) sovereigns. This is perhaps the problem with this new use of language: discourse; it is addressed to an internal 'population', but in the name of an external sovereignty supporting the performative authority (*speech act force*) of its enunciation. Or it addresses external sovereignty on behalf of the people, as in the ideal model of modern 'democracy'. The ambiguity of discursive enunciation is probably constitutive, in this sense.

3.2 Urban Chorematics

The instance of the city, abode of beings of stone, in our ecological terminology, involves another structural complexity to be studied. This is because, within the walls of the city, intramural, there must be three authorities: an administration, a place of government, temples and other sacred places (such as banks), and, between both, markets, places of exchange. In other words, the executive power, the sacred power, and, between the two, the effective, practical power of concrete values. The three authorities have this in common that they require continuous contact with the world, and first of all with the extramural territory, an effective means of both (1) administrative communication, (2) transport, especially food and of commodities, and (3) of doctrinaire control: in short, a transcendent dimension.

The cities therefore arise along the main roads, on rivers, on the shores of the sea, or if possible, on estuaries. For obvious reasons, the water crossing must allow each severed part to function fully. We can thus propose a *chorematic* model of the prototypical City; it embeds (swallowtail dynamics) within its walls a conflict of *choremes* with a clear intersection (butterfly dynamics); and it is marked by an ever-present transcendent cut. In the free space between its defensive walls and the three instances mentioned, one can note the existence of a public sphere, where conversation, art, entertainment, sport, "peripatetician" philosophy, poetry, erotic parade, etc., can unfold (Fig. 3.7).

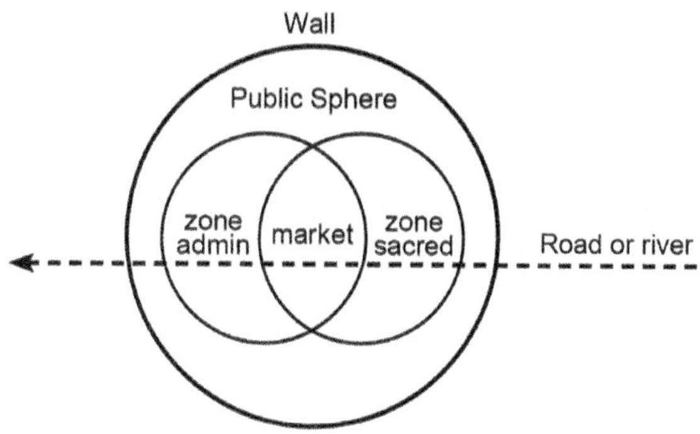

Fig. 3.7 *Chorematic* model of the minimal City

The visible and inhabited space of a city is obviously not reducible to such a geographical description; the model only describes a schematism which can constitute a kind of structure underlying what the plan of an existing, historical city shows us. A town hall, a covered market and a savings bank, a church…, and the remains of an old surrounding wall. If the categorical components are supposed to be present, they are obviously there historically, insofar as collective life has found them inscribed, projected, or projectable, on the geographical particularities of the place: river valley or main road, with hillocks capable of marking the areas to be distinguished. There is thus in general a semiotic *blending*[12] between the *chorematic* schematism and the territorial particularity which is offered to an urban development. The perception of what surrounds us is always double: we perceive the geographical horizon which is offered, and we perceive the *chorematic* distribution which makes the first have a meaning; we project the last perception onto the first, and we "*live in the blend*" of the two, as the cognitivist Mark Turner puts it. To understand this process of blending, it is necessary to go through the following explanation, which will study the relationship between *choremes* and mental spaces (see [9]). It is useful to start with the morpho-semantics of language.

3.3 Morphological Chorematics. Aspect, Time, Mode

Let us therefore return to the elementary phenomena of everyday and linguistic semantics. The form of the journey of the *choreme* S through the *choreme* P is expressed in the *aspectual values*, or in German *Aktionsarten*, which vary according to the languages and the stagings, but which do so in relation to a few particularly simple invariable principles. To begin is to enter, for an S, into a P. The *inchoative* therefore consists in crossing this *chorematic* threshold. This may require effort ("I *have finally* understood…", inchoative of insistence). The durative marks the fact of staying there, especially while something else happens, more or less abruptly, in *aorist* ("I was doing the dishes, when the telephone rang…"), so that the condition P also marks the new event that enters P ("So I lost the phone in the sink…"). The aspectual constellation of states and events is essential to their articulations. The *terminative* marks the exit, once again more or less difficult, which our adverbs willingly specify ("I *finally* quit smoking"). But the iterative adds a new entry or a whole series of new entries on the same *choreme*, which can condition what eventually happens there ("He was coughing all the time, so I didn't quite understand this he was trying to tell me"). Aspect is often used to bring the times of events into spatial contact. In visualization (Fig. 3.8) as in cognition, the trajectories of entities E on the realization of *choreme* R schematize the meaning of aspectual values (a trajectory can zigzag when E collides with the repeller of the *choreme* R).

[12] Semiotic blending, a concept developed by Brandt [1], is a meaningful application of one content to another content; we will discuss it technically in the following.

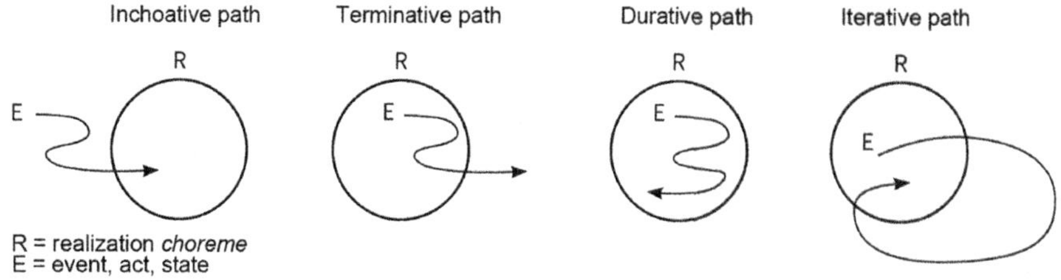

Fig. 3.8 Elementary aspect values

Fig. 3.9 *Still—already.* The drama of stress

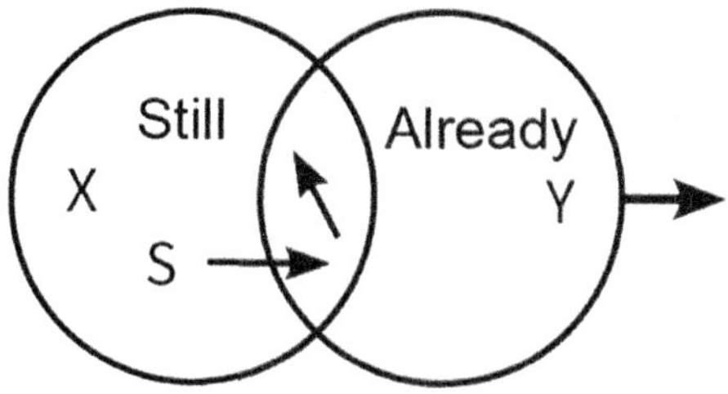

This is the case, in discursive semantics, of the particular conjuncture of "stress" or carpe *diem*. It brings together two aspects of transition, the terminative and the inchoative: we are *still* in a *choreme* from which we will soon emerge, but *already* in another *choreme* from which we will not escape for long. In the magnificent but cruel sonnet of the Spanish baroque poet Luis de Góngora, *Mientras por competir con tu cabello*, the lady learns from the poet's voice that she is *still* beautiful, but that she has *already* lost her youth, she must so let herself hasten to love (the poet), let her make up her mind while there is still time, when it is already late (Fig. 3.9).

The subject is still in X, but already in Y, from which he cannot leave. And Y moves away from X.

It can be an ecological or personal situation that is *still* livable (X), but *already* degraded (Y), and where the intersection X ∩ Y is destined to disappear: S will have lost condition X and will find himself exclusively in condition Y.

The *modalities* have to do with the consistency of the barrier, which can open and therefore let the object, the subject or the *choreme*, mobile, pass—enter or leave. Hence the importance of negations, a door (barrier) closes, and two negations appear, a door (barrier) opens, and the two negations disappear: **not** *being able* to be or enter X (e.g., a discotheque), **not** *being* able **not** *to be* in X or to leave it (e.g., a prison). We *must* then stay out or stay in. Whether it is the realization of a *being* or a *doing* makes no modal difference. *Must-X* always expresses a *May-X* under negation, either simple or double; in both cases, the realization barrier is closed—whether X is outside or inside.

One can also, in the intersubjective cognition that we call *communication*, schematize X as a relation to the other, which makes it possible to specify the modal regime, and thus to distinguish the *deontic* modality from the epistemic and the fundamental, *alethic*, which, in this intersubjective figure, is *performative*. The subject and the other are then considered as *choremes*, and the communication is seen or felt as a sharing and an exchange, which transports entities between the two personal *choremes*.

Indeed, when two people communicate, they "open up" to each other, and this double openness creates *possible* exchanges. We thus establish an intersection between the *choreme* of oneself[13] and the *choreme* of the other, a shared zone, where we can deposit themes or objects to be considered together, or more fundamentally negotiate the space occupied.[14] This makes communication possible, alethically; disagreement would immediately decrease the intersection, while total empathic agreement would increase it until it borders on merging. The intersection is a plastic attractor, of variable dimension.

The gesture of the handshake expresses the scheme. The ancient Greek custom of the *symbolon*, by which the friends each keep a piece of terracotta broken in two, refers to the intersection symbolically restored by the assembly of the two pieces after a separation. We can also quarrel, and start to become verbally aggressive: we then take advantage of the openness of the other to deposit negative objects in the shared area, which is called arguing *ad hominem*. In this way one bites, so to speak, into the flesh of the other; some do it literally, at particularly heated moments of the agonistic exchange. The transition to physical violence is always possible. In the quarrel, the intersection becomes a hand-to-hand fight by which one of the bodies tends to seize a part of the *chorematic* space of the other.

S2 invades S1's peripersonal space and eliminates objects: contempt, offense.

Intimate intersections move quickly to such stages; *chorematic* interpersonal openness is always dangerous. *Power* relations are established, precisely through the effect of the weakening of *chorematic* barriers—which are therefore psychological and social realities to be taken seriously.

If S2 in the communication graphs (Figs. 3.10 and 3.11) is this time a community, while S1 is a prospective member of this community, the relationship becomes *deontic*. S1 must (=*cannot not*) then honor the moral and often physical norms and behaviors of S2 in order to obtain his participatory identity passport. He has to pay his taxes, for example. He must renounce certain practices and adopt others in order to be able to register. This explains the opposition between *ethics* and *morals*: while *ethics* concerns respect S1-S2 from individual to individual, both being bodily and mortal, *morals* express the bond S1-{S2}, the tribute paid by the corporeal, mortal individual,

---

[13] We all project a small circular "peri-personal" space around us, literally, which decides the distances reserved for certain interactions, intimate, friendly, collegial, commercial, etc. We mark our attitudes towards the other by physically approaching or distancing ourselves from him, in everyday theatricality. This phenomenon is called *proxematics*.

[14] If I say to someone, "You can come in!", I give permission to the other to enter (in my office, for example), that is to say that his power to enter is created by the performative gesture of my saying. I then control the state of the door/barrier, i.e., the strength of this repeller which protects my (bureaucratic) *choreme*.

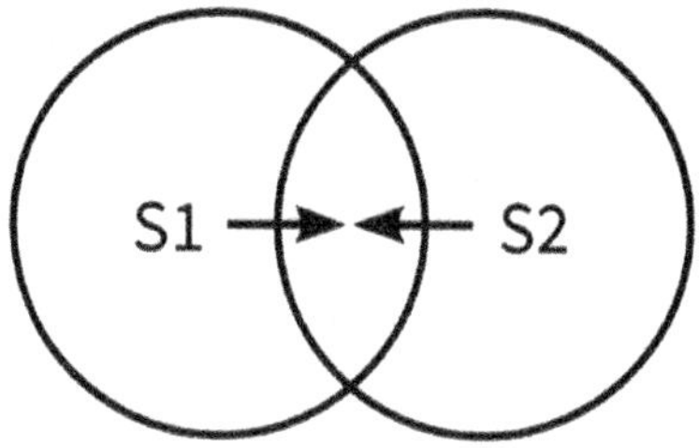

Fig. 3.10 The communication

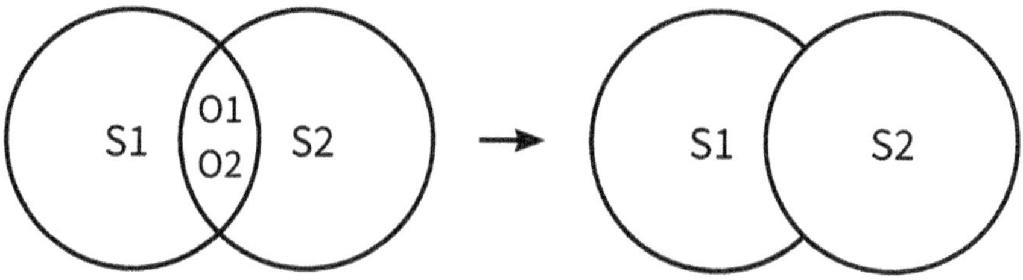

Fig. 3.11 Quarrel

to the abstract group and in principle immortal, and for all that "spiritual".[15] Deontics is a moral regime of modality.

Epistemic modality, on the other hand, is a *critical* device connecting two mental *choremes.* One contains a propositional idea, a postulate, an imaginary scenario, a hypothetical and simply thinkable state of affairs, in short a thought (P1); the other constitutes the set of thoughts ({P2}) which, according to the subject, deserve to figure in his knowledge and to remain in his memory. The critical question is whether P1 can be admitted into the set {P2} or not. This last set is a Truth *choreme* for the subject[16]; P1 must therefore correspond to certain criteria to be admitted, it must be 'likely' in relation to the prior content of {P2}, and it must offer features that meet the subject's expectations, hopes or fears, but also its aesthetic, scientific or philosophical meaning. In a sense, P1 must once again 'pay a price' to be admitted. We cannot believe in everything and anything. *Possible* ideas are admitted, often at the cost of reformulation, and the *impossible* ideas are rejected and must remain outside this inner forum containing what the subject judges to be able to 'admit', either as probable truths (which risk coming out of new), either as necessary truths, indispensable to the mental well-being of the subject, or to his philosophy.

[15] Spiritual simply means immortal, as is a group, a community, due to the fact of being composed of replaceable elements, and which are part of the group, by sharing its community "spirit".

[16] This Truth *choreme* is not necessarily organized as a vision of the world, a *Weltanschauung*, but can be a more or less chaotic and contradictory conglomeration, depending on the mental style of the subject.

Linguistic *modes* relate to this last point, to the mode of being of our contents of thought. In subordinate clauses, especially in the object complements of a transitive verb expressing a mental attitude, the mode manifests itself semantically with a certain clarity.[17] In the Roman languages, the variations between *indicative* and *subjunctive* play under these conditions on four values of verbal meaning expressing mental attitudes.[18]

We typically have two morphematic series, therefore two sets of 'signifiers' (indicative, subjunctive) for four signifieds that we will name α, β, γ, δ:

α *intentionalis*: I want/order that X: *subjunctive*;

β *imaginalis*: I imagine/fear that X: *indicative/subjunctive;* conditional;

γ *intellectus*: I know/see that X: *indicative*;

δ *affectus*: I regret that X: *subjunctive.* Indicative in Latin.

Italian, Portuguese and Spanish have more cases of subjunctive in β than French, due to the variable persistence of the Latin system, where the subjunctive is essential. Other languages and language families, without subjunctive morphology, use modal or syntactic expressions for the same values (α: I want this *to be* done today; β: I fear (that) he *will leave* me; δ: I regret that this *may* hurt you). The semantics of sentence embedding thus articulates a space–time ET1 where a subject envisages a scenario, and a space–time ET2 which would be that of this same scenario envisaged. In α and β, the content of ET2 is not asserted in ET1 as real, R, whereas it is asserted as real in γ and δ. On the other hand, in α and δ, the sense of ET2 is assumed in ET1 as *performative*, P, it modifies the relation between the subjects in ET1, whereas this is not the case in β and γ. The subjunctive therefore strongly expresses, in French, either the constellation ÷R & P, α, or the constellation R & P, δ. The neutral, constative, 'objective' attitude, in γ, corresponds to R & ÷ P; and the rather questioning, uncertain attitude of β is summed up in two negative criteria, ÷R & ÷P. In summary:

α: ÷R, P—ET2 is desired by the subject of ET1;

β: ÷R, ÷P—ET2 is just imagined by the subject of ET1;

γ: R, ÷P—ET2 is assumed as true by the subject of ET1;

δ: R, P—ET2 is true and affectively valued by the subject of ET1.

This RP morphology is obviously of the greatest importance for thought, since it is through it that we relate and refer to worldly and temporal reality, to the past, to the future and to the possible in general. that is, to everything that must be situated

[17] The mood manifests itself in many other types of subordinate clauses, and it must be seen in direct relation to the moods α–β–γ–δ in the main clause: *the imperative, the interrogative, the conclusive, the exclamatory.*

[18] I resume here my study on the question, "Mode, textuality", published in *Revue Romane*, VI, Fasc. 2, 1971. I use the same notation as in this article. The idea is developed in "It's five o'clock: Micro-prosody and Enunciation" [3] which studies detectable variations, even in sentence intonation.

outside the choreme of our present (ET1). We plan for the future (α) and evaluate it, when it has become a thing of the past (δ); we examine and interrogate the permanent and the recurrent (β) and we observe their transformations (γ).

Chorematically speaking, this object-completion mechanism of the grammar of (romance) languages shows us, by summarizing it, a universal phenomenon: the elementary epistemic functioning of the human mind. Among all the *choremes* we experience—the room, the house, the city, the country, the continent, the planet; groups, populations, cultures…—there is one and only one that we never leave: the phenomenal *choreme* of the *present*, the one that contains the present perception of what surrounds and inhabits us, of our own body, our attention, and our active imagination by which we project, thanks to the ET1—ET2 (supra) and the clutches[19] α, β, γ, δ, voluntarily or involuntarily space-times around us which *make our reality*. I will try to offer an analysis of this phenomenological part of the dynamics of meaning.

I let myself be guided by a personal example. I think of my apple tree. He is regrettably being smothered by an evil lime tree. So I promise myself to go and get involved in the vegetable fight and cut the linden branches that are hindering the proper development of my apples. The subject (me) actually uses the four shifters before deciding; he sees the real situation (γ), he regrets and reproaches himself (δ) for not having reacted sooner, he wonders what to do now (β) and he resolutely sets off with a ladder and a saw (α), because he wants those pernicious branches gone, and so he has a plan.

There is therefore a basic *choreme*, that of the subject, ET1, linked mentally, and possibly by grammatical expressions, to a series of mental spaces, ET2, that is to say to imaginary *choremes* temporally shifted, a constellation which finally leads to a decision taken in the basic *choreme*, to an *I-want* event. The result is a network which corresponds to a process by which three units of meaning, established by γ, δ and β, meet in a *choreme* of integration, often narrative, as in our example, ending up in the volitional plane leading back to the base ET1 by α. The contribution of β is decisive for the practical implementation, the schematic realization, of the call for change introduced by δ. Here is the processual network of the six semiotic instances involved in this small episode of practical thought. It coincides with the graph of conceptual integration known as the semio-cognitive model of 'blending' (see above), which integrates 'mental spaces', and this is no coincidence: the imaginary *chorems* (ET2) form a class of choremes specified by being anchored in a basic *choreme* (ET1) by a modal coupling (α/β/γ/δ), which precisely makes them imaginary, or mental, depending on terminology (Fig. 3.12).

The integration, here narrative, is done according to the inputs from the imaginary components of this decisional loop (note the direction of the arrows of projection, clutch, and perspective maturation). The model suggests that the production of meaning in such cases is somewhat circular, a looping process: it takes place by

[19] *Clutches* in the sense of Greimas, but in an inverted problematic, where it is a question of the fragile link between the 'bubble' of the subject and any other bubble. For Greimas, generative, the objective, mundane meaning was established before the subject entered the scene.

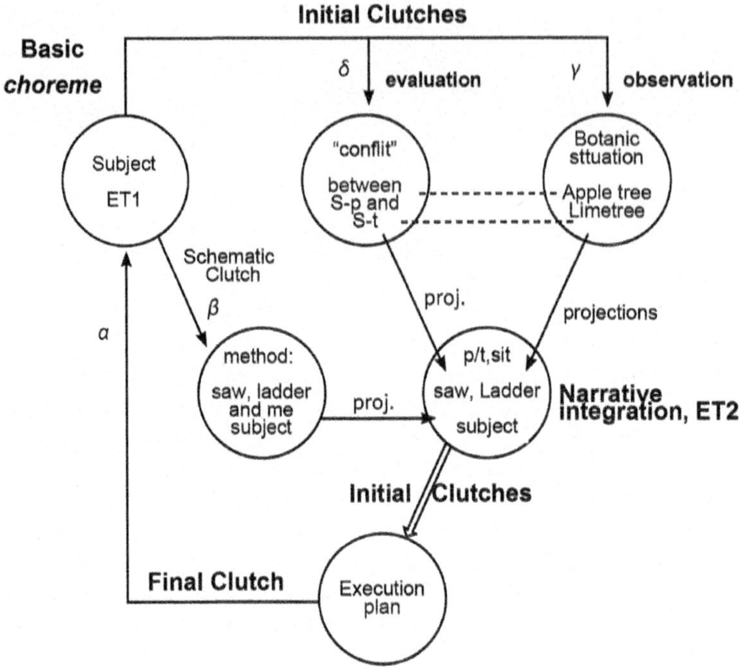

Fig. 3.12 *Chorematic* integration

efferent and then afferent clutches, in relation to the basic *choreme* (*basic space*). *Wanting* is indeed felt as a culmination, as the completion of a process of weighting, even if it is automatic; it is like a state that happens to the subject ("my decision is mine, because it comes to me, it *comes to my mind*", the subject could say). Ideas 'come' to us, they come to us.

Efferent attention is never volitional, it observes, it evaluates, or it dreams, but leaves the subject alone as to his participation; only volition engages the (imaginary) body of the subject or of his other ("I want you…"), the subject's delegate. Volition is afferent!

If this analysis is correct, thought therefore proceeds in a loop, when it combines and integrates sometimes abstract, sometimes concrete components, by regular returns to the basic *choreme*, where the stages of our often sinuous and undecided reasoning thus accumulate. It no doubt constitutes a perpetual 'flow', but this flow is modally organized, and it is probably this universal organization that explains why we generally manage to follow the reasoning of our co-subjects. There is a dynamic structure, a certain mental cuisine, which controls the vagaries of our content by channeling it into pre-established *chorematic* networks. Conceptual integration, blending, is a *chorematic* integration; this is probably one of the most important operations for the exercise of elementary intelligence.

3.4 Topological Thinking. Ontology

The analyzes presented illustrate, although punctually, the role that *chorematics* can play in bridging cognitive semantics, with its *containers* and its *force-barrier* schemes, and the structural and dynamic semiotics, topologically oriented, that which philosophical work -mathematics of René Thom inaugurated. Thom's catastrophic models, in particular the cuspoids, already make it possible to deploy the elementary topological spaces at work in the dynamic schematisms that we discover by studying cognition in vivo, which is the vocation of semiotics, following the phenomenology. The experience of the topographical situations where the *dramatis personae* and this true *genius loci* that is the place itself meet—the fury of the storm over our head, the rising waters, or the quicksands which engulf us, in short, the fundamental instability of the forces which envelop us and condition our acts[20]—this experience calls for a dynamic description of the relationship between things and their spaces of 'support', and in particular between these spaces. Thought itself is topographic and, as we wanted to show, topo-logical.[21] But the thought of thought understands this only if it looks at itself from the outside, which it manages to do to the extent—certainly unstable and fragile—to which it finds the diagrams at work when, spontaneously, it activates its schematisms and express them. If the semantics of language is topological, it is because it must simulate the 'semantics' of thought in order to be able to express it. And if the graphic diagrams describe the cognitive operations, it is because they simulate the mental diagrams of the thought itself. These simulations, we constantly criticize and correct them, which shows their contact with what they aim for. And this ontological observation is promising; thought can, at least in principle, understand itself. Otherwise, no semantics would be possible.

References

1. Brandt, L.: The Communicative Mind. A Linguistic Exploration of Conceptual Integration and Meaning Construction. Cambridge Scholars, Newcastle upon Tyne (2013)
2. Brandt, P.A.: Cognitive Semiotics. Signs, Mind and Meaning. Bloomsbury Advances in Semiotics, Londres (2020)
3. Brandt, P.A.: The Music of Meaning. Essays in Cognitive Semiotics. Cambridge Scholars, Newcastle upon Tyne (2019)
4. Brandt, P.A.: La charpente modale du sens. Benjamins, Amsterdam (1992)
5. Brandt, Per Aage (1982). Sandheden, sætningen og døden. [La vérité, la phrase, la mort]. Copenhague: Basilisk
6. Brunet, R.: La composition des modèles dans l'analyse spatiale. L'Espace géographique, n° 4 (1980)
7. Derrida, J.: Khôra. Galilée, Paris (1993)

[20] Favourite idea of Michel Serres, see *Genèse* [12].

[21] One could say that formal logic offers a portrait of thought deprived of space–time, a deprivation which extracts it from the imaginary, the only place where it can invent and make synthetic discoveries, and not only analytical ones, in the sense of Kant and analytic philosophy.

8. Greimas, A.J., Courtés, J.: Sémiotique: Dictionnaire raisonné de la théorie du langage. Hachette Université, Paris (1979)
9. Fauconnier, G., Turner, M.: The Way We Think. Conceptual Blending and the Mind's Hidden Complexities. Basic Books, New York (2002)
10. Kristeva, J.: La révolution du langage poétique. Editions du Seuil, Paris (1974)
11. Lakoff, G., Johnson, M.: Metaphors We Live By. Chicago University Press, Chicago (1980)
12. Serres, M.: Genèse. Grasset, Paris (1982)
13. Sweetser, E.: From etymology to pragmatics. Cambridge University Press, Cambridge (1990)
14. Talmy, L.: Toward a Cognitive Semantics, I et II. En particulier le chap. 7: Force Dynamics in Language and Cognition. MIT Press, Cambridge, Mass (2000)
15. Thom, R.: Apologie du logos. Hachette, Paris (1990)
16. Thom, R.: Stabilité structurelle et morphogenèse. Essais d'une théorie générale des modèles. W. A. Benjamins, Reading, Mass (1972)

Chapter 4
Topological and Dynamic Morphologies in Semiotics

Wolfgang Wildgen

4.1 Catastrophe Theory and Its Applications in Semiotics in Their Historical and Contemporary Context

Since Plato's dialogue, "Timaeus", in the 5th century B.C.,[1] attempts have been made to use mathematics to explain the principles of nature, the human spirit, and culture. For Plato, the geometry of Pythagoras, that is to say, the right-angled triangle, the regular polygons, and the regular polyhedra (Platonic solids) served as a model for the creation and composition of the universe and the mind. His disciple Aristotle preferred to use logical schemes (categories, syllogisms) to postulate a structure of thought, language, and scientific discourse. These two currents had successors for two millennia. Already in antiquity, the art of rhetorics contained a component of "memoria" i.e., the art of memorizing the contents of a public performance using a spatial organization of these contents. The "art of memory" exploits the cognitive faculties of our orientation in space and the geometry/topology of a journey in this space to optimize the activation of the natural mnemonic capacities.[2] The "theatre of memory" was even concretized in Renaissance architecture. This technique proved the geometric/topological nature of the semantic space. Words are associated with images, under certain conditions, with musical or olfactory forms. In the 16th century, the image spaces of Giordano Bruno and his art of memory were confronted

[1] This article was originally a contribution to the research seminar: Actualité de René Thom, organized by Isabel Marcos et Clément Morier at the university Paris-Diderot. I have translated the text into English and added the Sect. 4.3 hat summarizes an argument in Wildgen [41].

[2] Topology is a generalization of geometry and geometry is the particular case of a topology provided with a metric.

W. Wildgen (✉)
University of Bremen, Bremen, Germany
e-mail: wildgen@uni-bremen.de

© The Author(s), under exclusive license to Springer Nature Switzerland AG 2024

I. Marcos and C. Morier (eds.), *The Relevance of René Thom*, Lecture Notes in Morphogenesis, https://doi.org/10.1007/978-3-031-54982-3_4

with the logico-semantic hierarchies of Pierre de la Ramée.[3] In the 17th century, geometrical and dynamical models gained a new impetus when Galileo and Kepler began to treat planets' movements in astronomy, projectiles in ballistics, and the movement of animate beings with the same mathematical tools. Newton and Leibniz introduced differential calculus, and Euler developed the mathematical means to deal with many dynamical problems, from physics to music. At the beginning of the 19th century, Wolfgang Goethe introduced the notion of "general morphology" in the context of his research in botany (the "Urpflanze") and zoology (the skeleton and the morphological variations of the "original vertebra"). August Schleicher used his notion of "morphology" to study the structure of words in Indo-European languages, and Wilhelm Humboldt in his typology of languages.[4] However, morphologies exist not only in their spatial extension. They are dynamic, vary over time, unfold, change in predictable and calculable ways, or remain stable through forces that eliminate all minor fluctuations. Towards the end of the 19th and the beginning of the 20th century, Poincaré, and later researchers in the theory of dynamical systems like René Thom, gave an exact form to the theories of structural stability and classified the universal unfoldings of non-stable but simple dynamical systems. Since the 1970s, the study of dynamic systems has grown exponentially with chaos theory, fractal geometries, and stochastic dynamics.

In the same period of innovation in topology and analysis, i.e., from 1870 to 1970, a "zoo" of logical devices appeared, of which we must mention the work of Peirce in his logic of relations around 1900, the contributions of Whitehead and Russel after 1910 ("Principia mathematica"), the theory of automata and programming languages, the intensional logic of Montague, the logic of situations of Barwise and Perry, and further variants. Some of these innovations have led to modeling in linguistics (indirectly in semiotics inspired by linguistics). René Thom was the only one to consider the tradition from Plato, renewed in Renaissance Platonism (Ficino, Bruno), applied to botany and zoology by Goethe, and philosophically in the "Naturphilosophie" of Schelling. It found its modern form in applied (so-called qualitative) differential topology and catastrophe theory. The applications of Thom's semiotic topology are a kind of Robinson's island in the Gulf Stream of Chomskyan and logic-based linguistics. The Chomskyan enterprise has succeeded in establishing a linguistic ideology that controls the criteria of relevance and the descriptive and explanatory goals of the discipline in such a way as to downgrade all rival enterprises. The critics of his models hardly sought to overcome the narrowness of the formal choices; they rather rejected any mathematization, or they used more primitive formal means such as premodern logic and mathematics (see the role of the semiotic square in Greimas,[5] the "natural logic" of Lakoff and later his (cognitive) semantics of metaphors).[6]

---

[3] See Wildgen [52] for Bruno's artificial memory theories and Wildgen ([40]:171–186) for Bruno and Ramée.

[4] See Wildgen ([48]: 239–241) for the contribution of Goethe and that of Humboldt (ibidem: 246s.). Humboldt speaks of the "inner form" ("innere Sprachform") of languages.

[5] See the review by Petitot [21].

[6] This argumentative figure is, moreover, old. Thus the innovations of the "Jung-Grammatiker" of Leipzig of the 1870s were rejected by linguists of the historical stages of languages. After the

Chomsky had started a program of mathematization in his 1955 thesis; consecutive variants of his model reduced the mathematical aspects to give more room for the adaptation to empirical requirements. In a way, they responded to descriptive demands introduced in American descriptivism (see Bloomfield, Pike, and Harris) and abandoned the earlier goal of a mathematically strict theory.[7] Chomsky hardly reacted to the criticism of the formalism of transformations (too general) and the abandonment of statistical aspects (considered basic by Harris). After 1980, artificial intelligence turned to "neural nets" and against programming based on explicit rules. It can be concluded that the mathematization within linguistics has led, on the one hand, to very ritualized models with an increasingly poor mathematical structure. On the other hand, it has hardly followed the development in the specific domains of mathematics and their advanced applications in the natural sciences.

This semi- or quasi-formal tendency is also characteristic of the work of Lakoff and Langacker in cognitive semantics. They borrow many concepts from gestalt psychology or cognitive science to achieve a quasi-pictorial schematization and reduce the formal tools to very elementary logical means (for example, networks of relations or graphs). Langacker's critics deplored his lack of systematicity and the purely didactic character of his image semantics and cognitivism.[8] A similar strategy is employed by Greimas and his successors when they define the semiotic square or the networks of semes and isotopic relations. These notions refer to a logic that is never explicitly formulated and whose technical possibilities are not exploited. As a result, the mathematization remains incomplete, if not contradictory.[9]

The classic example of a sound mathematization in semiotics is given by Charles Sanders Peirce (1839–1914), himself a logician and chemist and an expert in the philosophy of science. He conceived a logic of relations and the art of thinking/arguing clearly and exactly (in the spirit of Descartes' "idées claires"). Topological

---

victory of this innovative current and its establishment around 1900, the innovative ideas of Ferdinand de Saussure were ignored by linguists in the tradition of comparative studies. However, they could not slow down the establishment of the structuralist movement of Jakobson and Hjelmslev, which became the new standard for the schools of Greimas and Eco in the years after 1960. The Chomskyans, in turn, tried to establish a new standard of scientificity, aided by the rise of computer science and artificial intelligence.

[7] The Chomskyan movement even succeeded in destroying the advances in mathematical linguistics of its teacher Harris, who was much more open (and critical) towards the mathematical techniques to be employed in linguistics; see the chapter "Harris und Chomsky" in Wildgen ([57]: 94–100).

[8] See Wildgen ([51]: chapter 1) for a critical review of his and comparable research. In Wildgen [56], the whole field of "Cognitive Grammar" is reviewed (in German).

[9] See the article: "semiotic square" by Jean Petitot (in Greimas and Courtès [12]). As this strategy is tempting for disciplines that do not have a stable tradition of mathematization, we can expect that it will also appear in the field of CT applications. This was the core of the criticisms of certain works of Christopher Zeeman by Sussmann and Zahler [30] and can probably be extended to some of the applications of CT in psychology (psychoanalysis), history and semiotics. In the epilogue to his 1980 book, Gilmore refers to the controversy of the 1977/78 as "tempest in a teapot" and he reduces this discussion to a small number of questions to which the TC can find an adequate answer (see Gilmore [11]: 643–646).

aspects (a mathematical discipline that developed during Peirce's career) were integrated into his logic of diagrams. He also sketched out a topological reading of verbal valences. In this sense, he can be considered the precursor of René Thom's topological semantics. The relational nature of the grammar of natural languages and the prominent role of valence was also at the center of the structural syntax of Tesnière (1893–1954), who developed this theoretical approach in the 1930s (published in 1959). Peirce's work was unknown to him, however.[10] René Thom took Tesnière's linguistic intuitions as a starting point for his suggestions for morphodynamic linguistics [31].

In the 1960s, Chomsky's thought and approach to a generative and transformational grammar began to dominate theorizing in linguistics; see his "Syntactic Structures," published in 1957, and the standard model published in 1965, "Aspects of the Theory of Syntax" [8–10]. Chomsky, who had published papers in logic (in the early 1950s), was not a mathematician, and his school soon neglected the mathematical aspects. In the sequel, linguistic engineering followed its path favoring simpler algorithms, such as augmented transition networks (ATN), and statistical approaches, such as PDP (parallel distributed processes). The models proposed by the Chomsky school ended up ritualizing and diversifying into many variants with an increasingly narrow theoretical scope. This theoretical landscape, the deconstructivist (and relativist, nominalist) tendencies, and the philological conservatism of all those who refused Chomskyan "innovations" created an intellectual atmosphere unfavorable to René Thom's suggestions.[11]

These historiographic remarks outlined the scientific context of the work of René Thom when he turned to biology and the human sciences at the end of the 1960s[12] and of his development until the end of the last century. It must be seen that, unlike the undertakings of Hjelmslev, Greimas, and others, he started from a skill in mathematics and the application of mathematical tools in the natural sciences (especially physics). He contributed to the development of differential topology, earning him the Fields Medal (corresponding to a Nobel Prize) in 1958.[13] After settling at IHÉS in Bures-sur-Yvette (Institute of High Studies in Science, near Paris), Thom began a second career in the philosophy of science. Finally, he developed deeper and more thorough models in the humanities, including semiotics.

His intellectual partner in the 1970s in Europe was the English mathematician Sir Christopher Zeeman (1925–2016). In 1976 Zeeman published a dynamic model of cell differentiation in animal morphogenesis, which was empirically substantiated

---

[10] It was not until 1935, especially following the publications of Charles William Morris (1902–1979), that Peirce's works were published as a whole and known in the United States and Europe.

[11] See also the critical remarks of Thom concerning contemporary linguistics in his preface to Wildgen [53] and the commentary of his critique in Wildgen [43].

[12] Cf. [35, 37] and for an overview on the first writings of Thom on semiotics and linguistics Petitot [25].

[13] This is, moreover, a central feature of scientific "revolutions" of this kind in the 20th century, whose advocates were part of the cream of scientists and belong to the same generation (born between 1917 and 1927): René Thom (1923–2002), medal Fields 1958; Ilya Prigogine (1917–2003), Nobel Prize in Chemistry in 1977; Benoit Mandelbrot (1924–2010), polymath and "fractalist", and Hermann Haken (born in 1927), eminent physicist.

in 2003.[14] In addition, he helped spread ideas for models using catastrophe theory in the 1970s.[15] Finally, in 1982, a congress at the interdisciplinary center of Cérisy (France, Normandie) brought together the major representatives of this movement with the two leaders René Thom and Christopher Zeeman.[16]

Zeeman and Thom were not the only ones to seek new paths for mathematical theorizing in the natural and human sciences. However, they were the first to attract the attention and support of an international scientific public.

- Ilya Prigogine (1917–2003), whose work in chemistry earned him a Nobel Prize in 1977, was known for his work on thermodynamics and dissipative structures. His team in Brussels proposed applications in biology (see the phenomenon of self-organization of living systems and Prigogine and Stengers [28]). He also triggered an epistemological discussion in the natural sciences (including a controversy with René Thom).
- Benoit Mandelbrot (1924–2010), mathematician and computer scientist, proposed fractal and self-similar structures (without scale) as a constituent of many natural phenomena of geography up to the statistical modeling of the evolution of stock market prices (see for an overview Mandelbrot [18]).
- Hermann Haken (born in 1927) was a mathematician and physicist who generalized laser theory in an interdisciplinary approach called "synergetics".[17] In a series of conferences in Elmau (Germany), he brought together many researchers in the most diverse disciplines around his program.[18] An American school around Scott Kelso continued this work in a psycho- and neurological context (see Kelso [16]).

We can say that René Thom and Christopher Zeeman initiated interdisciplinary research enterprises based on modern mathematics and using the methodological and empirical standards of the natural sciences. However, as all the founding personalities of the movement belong to the generation from 1917 to 1927 and are dead or have reached the age of 96 (Haken), their initiative has lost momentum. Therefore, it is necessary to go beyond their approaches, responding to developments after 1980, sustaining the successes achieved, and responding productively to criticism. Therefore, in the next section, we go back to the beginning of Thom's enterprise in his provocative book: "Structural Stability and Morphogenesis" of 1972 (English translation in 1975, and second enlarged and corrected, French edition in

[14] See: https://en.wikipedia.org/wiki/Clock_and_wavefront_model.

[15] See the articles collected in Zeeman [65] and the introduction to this type of interdisciplinary work by Poston and Stewart [27].

[16] See the publication of the proceedings in Petitot [19].

[17] See: Introduction to synergetics, published in 1973 in German, in 1983 in English.

[18] In 2008, Hermann Haken published the book: Brain Dynamics [15]. I have contributed to two synergetics conferences in Elmau in 1989; see Wildgen [50], and in Bremen in 1994, see Wildgen [44].

1977) and comment on the evolution of his ideas for semiotics and morphodynamic linguistics.[19]

4.2 Morphological Principles as Deep Structures in the Lexicon of Verbs and Basic Syntax

Based on state of the art in differential topology, Thom stated the conjecture of a universal classification of the unfoldings of a structurally unstable germ, which led to the notion of elementary catastrophes in Thom [36].[20] Catastrophe theory is the application of this classification theorem. Today we prefer to speak of mathematical results in differential topology, especially in the theory of bifurcations and singularities. Catastrophe theory is representative of applied mathematics in this field. The starting point for the applications proposed by René Thom was Waddington's biology and its morphogenetic landscapes (see Waddington [38] and Waddington's foreword in Thom [36]: xii–xv). Like, in the 19th century, Johann Wolfgang Goethe (1749–1832) in his "Morphologie überhaupt" and Étienne Geoffroy Saint-Hilaire (1772–1844) in embryology and comparative anatomy, René Thom assumed a more general scope of morphogenetic principles. He saw linguistics as an "exemplary morphological discipline" [32]. In an article from 1970, summarized in his book "Structural stability and morphogenesis", he offers a list of archetypal patterns derived from elementary catastrophes. It was expanded, based on research published after Thom's book, in Wildgen ([45]; Habilitation thesis see partial publications in Wildgen [47, 49]). In the following sections, we shall give examples of this formalization and its applications adapted to a readership of non-mathematicians.

The years 1979–1990 saw several major contributions to applied catastrophe theory in semiotics (the thesis of Jean Petitot in [22] and Bandt in 1987) and linguistics (the thesis of Wolfgang Wildgen in 1979). Many articles discussed further details on this subject during this period. While Jean Petitot elaborated on the underlying epistemology and sought a link with the semiotics of Greimas (Semiotic School of Paris), Brandt [5] clarified the theory of modalities introduced by Greimas, using the concepts of catastrophe theory (C.T.). In the second phase (from 1990 to 2000), Petitot sought a link with the cognitive semantics sketched by Langacker, Talmy, and Lakoff and with mathematized neurosciences (see his neurogeometry developed in Petitot [24]). Brandt elaborated on the cognitive semantics initiated by Talmy and

[19] The English translation has dramatically reduced the chapter on applications in linguistics, which can be interpreted as an implicit criticism of this "Violin of Ingres" by René Thom. The second French edition (from 1977) corrected a certain number of details concerning the umbilicals given new results in mathematics, without however, modifying the list of semantic archetypes and without using the new research on the unfoldings called "exceptional" (E_6, E_7, and E_8) and their compactification in the double cusp X_9. The necessary corrections were the subject of a chapter in Wildgen ([47]: 81–92 and [49]: 201–222).

[20] An overview on the larger mathematical field linked to catastrophe theory is given by Arnol'd [2].

proposed applications of the theory of cognitive "blending" (see Brandt [6]). We will limit the exposition to the results of our research in this period.

Semantic archetypes are, first of all, paths in the critical space of a catastrophe. We shall specify the case of the elementary catastrophe called the cusp. As shown by Guckenheimer and Holmes [13], the problem of classifying paths in the bifurcation set is still straightforward for the cusp, where this set has two dimensions (u, v; control variables). For catastrophes with 3, 4, 5, ..., n control variables, it is less obvious, and for the umbilics, we do not yet have the necessary means to derive all the four-attractor schemes proposed by René Thom in 1972. We must go beyond elementary catastrophes; for instance, consider so-called "exceptional" catastrophes (E_6, E_7, and E_8) and their compact variants embedded in the double cusp, X_9.[21] In response to the criticisms of Sussman and Zahler [30], one must be very explicit in the interpretation of internal variables (fast dynamics) and external variables (slow dynamics). The standards observed in applied mathematics in the exact sciences also apply to "soft" applications in the human sciences. The work in morphodynamic linguistics by Wildgen [46, 47, 49, 51, 53] elaborated Thom's proposals for linguistics and semiotics and confronted them with the results obtained in empirical linguistics (see for recent contributions Wildgen [61] and [41]: chapter 6).

To understand the relevance of C.T. models in linguistics, one must first understand the continuity of morphogenetic programs from biology (from the cell to the whole body) to the behavior of man (and animal) and their means of communication, especially man'/woman's means of symbolization, including the use of language. According to Thom, the common denominator of all these phenomena is the spatial and temporal organization, whose archetypes are reflected by the differential topology, the singularities, the bifurcations, and the stable unfoldings. The list of catastrophes gives the core of this morphogenetic "zoo", which can be enriched if we consider generalized catastrophes (which in the 1970s were called strange attractors, chaotic developments). We can add the general laws in Prigogine's analysis of "dissipative structures far from thermodynamic equilibrium" and "slaving parameters" described by Hermann Haken [14]. The analysis assumes mimesis of the spatio-temporal organization of living things in social communication and language. It clearly shows itself at the sentential level in the architecture of verbal valences and the narrative schemas of discourse. It is only against the background of this morphogenetic theorizing that C.T. models make sense in linguistics. Suppose we restrict linguistics to a purely descriptive (or even normative) program of enumerating existing patterns, segmentations, and sequential regularities of some historically

[21] René Thom had postulated a limit of external variables (called external or control variables) at $n \leq 4$, i.e. corresponding to the dimensions of space-time. I prefer to neglect this restriction which would apply only to interpretations of concrete movements in space-time. The notion of "simple" catastrophe in Arnold [2] can replace this restriction. However, it allows a countable infinity of attractors for cuspoïdes. A more general limitation applies to umbilics and exceptionals. This limit corresponds to the limit of the Platonic polyhedra (to five or to three, if we consider the two dual polyhedra: cube $\approx$ octahedron and dodecahedron $\approx$ icosahedron). See for this parallel Slodowy [29] and Wildgen [63].

documented languages. In that case, such a project is not of major interest.[22] The morphogenetic view also bans any nativist or creationist hypothesis.

4.2.1 *The Geometry and Dynamics of Elementary Catastrophes: The Example of the Cusp*

We shall show how it is possible to classify paths in the cusp and how this gives rise to a list of morphologies, which Thom calls "archetypal morphologies". On the one hand, we shall keep contact with mathematical operations and, on the other hand, reduce the technicality necessary for explicit derivation and demonstration. The normal form of the cusp is the prototype of a large number of forms structurally stable under deformation, and one can add an unlimited number of quadratic functions to it. Moreover, the geometry of the normal forms stands for a much richer topology.[23]

$$\text{Potential function} : \mathrm{V} = \mathrm{x}^4/4 (\text{this variant simplifies the calculations}). \tag{4.1}$$

$$\text{Universal unfolding} : \mathrm{V} = \mathrm{x}^4/4 + \mathrm{ux}^2/2 + \mathrm{vx} \tag{4.2}$$

$$\text{Partial derivative} : \mathrm{V}' = \mathrm{x}^3 + \mathrm{ux} + \mathrm{v} \tag{4.3}$$

$$\mathrm{V}' = \mathrm{x}^3 + \mathrm{ux} + \mathrm{v} = 0 \tag{4.4}$$

$$\mathrm{V}'' = 3\mathrm{x}^2 + \mathrm{u} = 0 \tag{4.5}$$

The unfolding is difficult to represent by a graph because it has four dimensions. In order to know the critical points, we consider the zeros of the first (partial) derivative and the zeros (degenerate points) of the second derivative:

Equations (4.4) and (4.5) can be used to calculate the effect of the external variables u and v, also called control variables. We will perform it explicitly to show that this is a simple calculation.

In Eq. (4.6), we isolate the variable x from Eq. (4.5).

---

[22] In Chomsky's terminology, we could say that the explanatory adequacy is the major goal and not the descriptive adequacy. However, explanation has a different scope in a linguistic theory that considers morphogenesis and its biological roots as essential. In Chomsky's word view, such questions are short-circuited via his nativist hypothesis. Instead, considerations of formal simplicity must fill the explanatory lacuna.

[23] Detailed treatments can be found in Bruter ([7], vol. 1: 198–201; in French) and in Gilmore [11]: 97–104; in English).

$$x^2 = -u/3 \tag{4.6}$$

Equation (4.6) is used to replace x^2 in Eq. (4.4). This operation results in Eq. (4.7) at the right.

$$x^3 + ux + v = 0 \Leftrightarrow x(x^2 + u) = -v \tag{4.7}$$

$x(-u/3 + u) = -v \Leftrightarrow x(2/3u) = -v$. If this equation is squared, we obtain (4.8)

$$x^2(4/9u^2) = v^2 \tag{4.9}$$

$$(-u/3)\,(4/9u^2) = v^2$$
$$\text{(the value for } x^2 \text{in equation (4.6)is used again)} \Leftrightarrow$$
$$-4u^3/27 = v^2 \tag{4.10}$$

$$27v^2 + 4u^3 = 0 \tag{4.11}$$

This equation describes the bifurcation set, i.e., the space reduced to the control variables u and v. It describes the effect of the spatio-temporal process that underlies the dynamics in the scope of an application. The correspondent graph is a classic curve (a type of semi-cubic curve). We will give the graphs of Eqs. (4.4), (4.5), and (4.11) in the following figures.

The graph of the equation $x^3 + ux + v = 0$ gives a surface in the space (x, u, v), i.e., in the three-dimensional space of real numbers ($\mathbb{R}$ 3), which is smooth for values $u > 0$ but which has a double fold for values $u < 0$ that originates in the singularity $(u = v = 0)$.

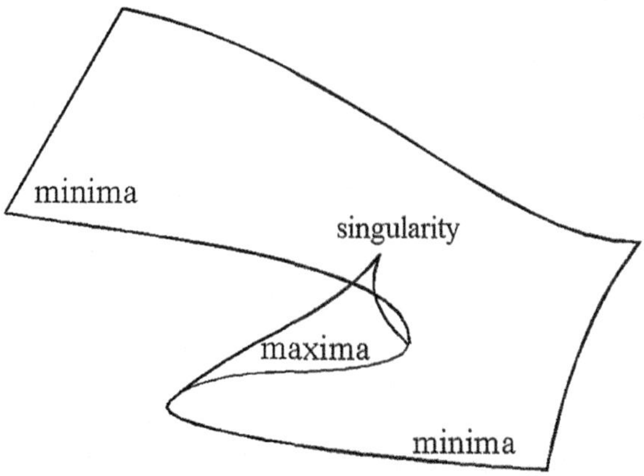

Fig. 4.1 The graph of Eq. (4.4)

The graph of Eq. (4.5) describes the contours of the double fold in the plane (x, u) in a horizontal projection of Fig. 4.1.

The graph of the semi-cubic parabola corresponds to a vertical projection of Fig. 4.1 in the plane (u, v) (Fig. 4.3).

We can represent the essential information contained in Eq. (4.2) by specifying, for certain points of the plane (u, v), the form of the potential V (in the plane V, x).

Fig. 4.2 The graph of Eq. (4.5)

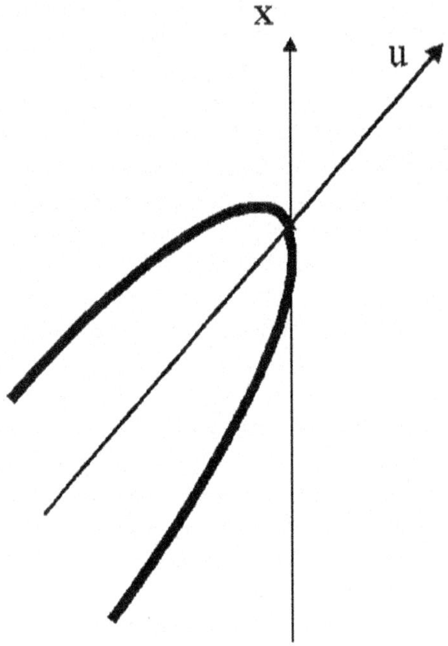

Fig. 4.3 The graph of Eq. (4.11)

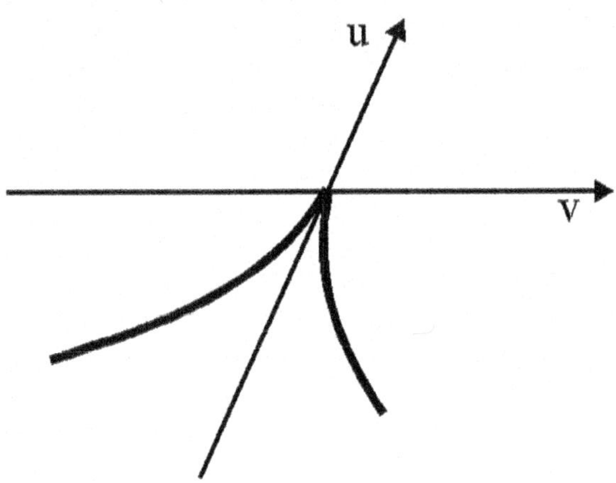

Figure 4.4 gives such a description and indicates a segmentation of the (u, v) plane into qualitative domains (Fig. 4.5).

Finally, the dynamic information of the cusp can be described by considering different sections of the surface of the critical points (maxima, minima, and halting points) and the (vectorial) orbits on x. Figure 4.6 gives two sections as the starting point for constructing a morphological diagram. The diagram at the right shows only the minima and the bifurcation points.

After having illustrated the geometry and the dynamics of the cusp in detail, we can consider a minimal diagrammatic representation of the information in a set of

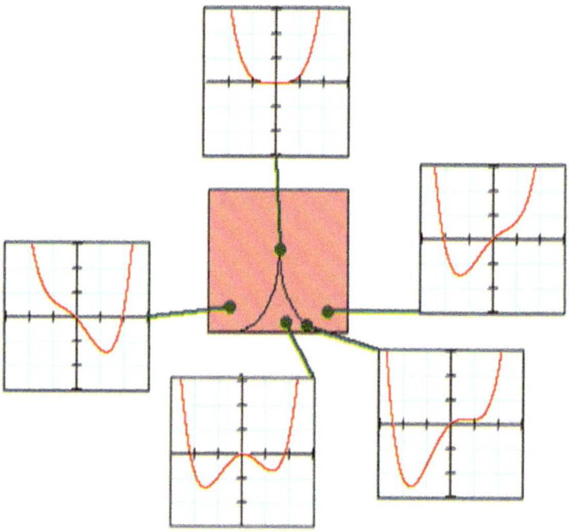

Fig. 4.4 The potentials of Eq. (4.2) for a choice of points in the (u, v) plane described by Eq. (4.11)

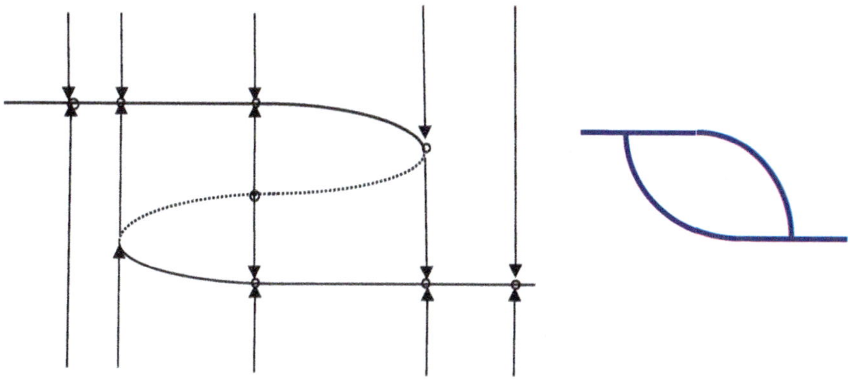

Fig. 4.5 Path following the variable v (increasing) on the surface of the critical points of the cusp and the corresponding simplified diagram (without the maxima)

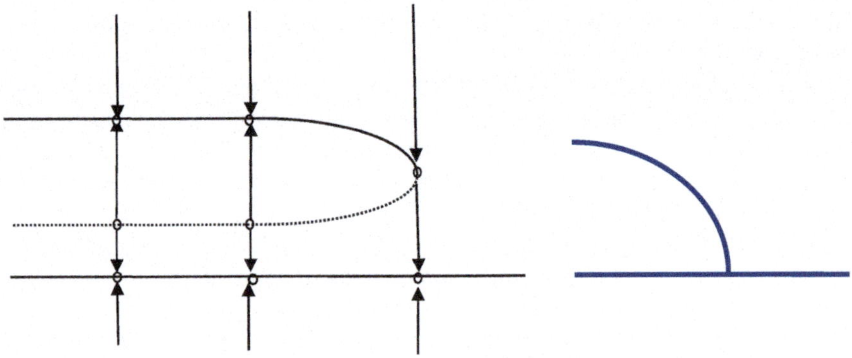

Fig. 4.6 Path following the variable u (increasing) in the surface of the critical points of the cusp and the corresponding simplified diagram (without the maxima)

catastrophes. This simplification will allow us to characterize the other elementary catastrophes and the set of archetypes derived from them. Our applications will, however, only consider the subfield of archetypes derived from the cusp; cf. Wildgen ([47]; in English, and [49]; in German) for further details.

The maximal information of a catastrophe, from which one can deduce the other features, is given by the zone containing the maximum number of orbits. If we go back to Fig. 4.6, we see in the left section two vector fields, each with three critical points (o) consisting of two attractors and one repeller; four non-zero vectors (positive and negative) describe the force fields that join them. This information can be represented by a diagram in which the positive signature designates the minima and the negative signature the maxima. Figure 4.7 represents, with this technique, the maximal sequence of extrema in the elementary catastrophe called "butterfly" (A_5: $V = x^6$, unfolding $V = x^6 + ux^4 + vx^3 + wx^2 + tx$). At the top, we see the graph of the potential (V) and the internal variable (x), and at the bottom, the type of the extrema (minimum or maximum) and the direction of the vector field.

As the vector field always goes from (−) to (+), a diagram without vectors can represent the critical points (+) and (−) and the orbits between these critical points. This diagram is called a Dynkin diagram. It has one dimension for the cuspoids and two dimensions for the umbilics. The singularities A_3, A_4, A_5, and A_6 and the duals (negative variants) of A_3 and A_5 can be characterized by Dynkin diagrams. We have chosen the maximal structure points as representatives. The duals of A_3 and A_5 are placed on the right side (Fig. 4.8).

4.2.2 *Actantial Schematizations Derived from the Cusp*

For concrete modeling, for instance, in linguistics and, in our case, for verbal valence, it is necessary to provide an interpretation framework for the variables of the formal

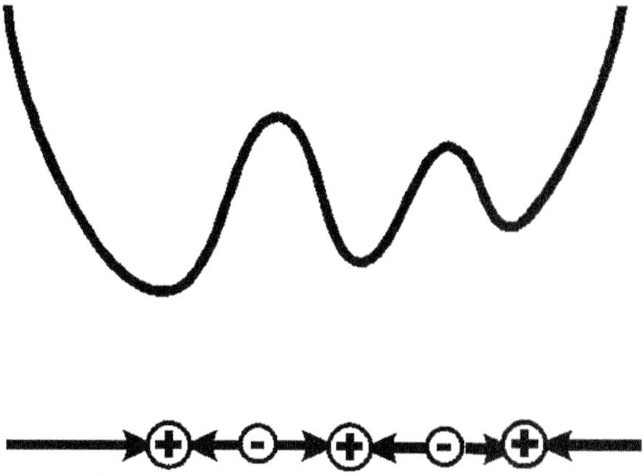

Fig. 4.7 The maximum sequence in the butterfly (with vectors)

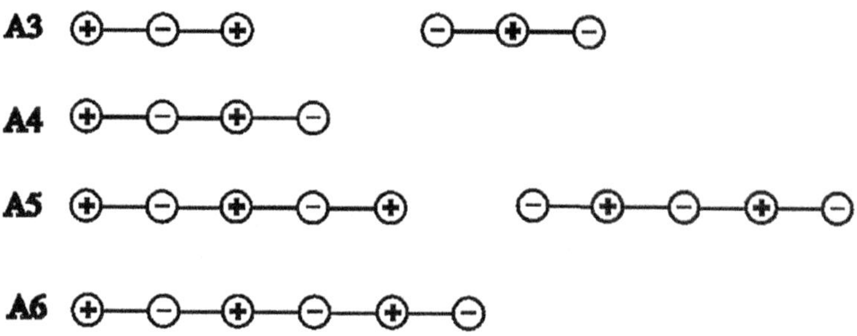

Fig. 4.8 Dynkin diagrams for some cuspoïds (the list of cuspoïds is denumerable infinite)

model. In Wildgen ([51]: chapter 5), we proposed a hierarchy of "representation spaces". We will, in this contribution, choose one position in this hierarchy. Next, as a reference corpus, we chose the lexicon of German verbs that Ballmer and Brennenstuhl [4] established according to an intuitive classification of the underlying events or processes. This classification is independent of the catastrophist modeling and may serve as an empirical control for our approach. Finally, we give a catastrophist modeling for three groups of verbs distinguished in the lexicon of German verbs mentioned (for a complete analysis, see Wildgen [51]: chapter 3: The meaning of verbs, pp. 57–74).

In the intentional act of an *agent* directed at a less intentional *object* or entity, two aspects can be distinguished:

(a) The configurational aspect describes the topological and kinematic relations between the agent and its object.
(b) The energetic (or intentional) aspect introduces the force of the agent and the effect of this force on the object. This force is first of a psychic nature; it then extends to physical effects.

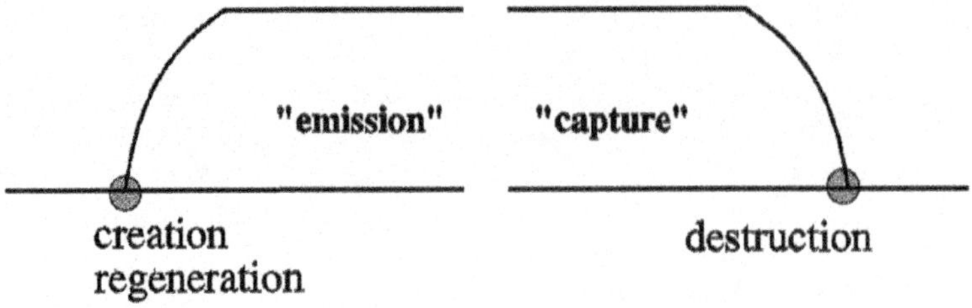

Fig. 4.9 The emission and capture archetypes

In their classification of the lexicon of German verbs, [4] distinguish, at this level of complexity, two groups of verbs:

(1) The creation, destruction, and regeneration of objects (elements of the environment)
(2) The impact of the subject on the state of objects and subjects in its environment

4.2.2.1 First Group of Verbs (Creation/Destruction)

For group (1), it is easy to see the correspondence to the emission and capture archetypes. Figure 4.9 shows the graph of the corresponding archetype.

In the lexicon of verbs, the corresponding verbs are, in most cases, divalent, as in:

Albert tells a story.

Berthe calculates the result.

Charles eats the soup.

These and further examples show that:

(a) The actantial structure of the verb does not correspond directly to the underlying schematization.
(b) The form of the lexicon and a fortiori of the syntax does not depend exclusively on the cognitive classification of process types; the distribution of processual information in the morphological or the syntactic construction has many degrees of freedom. Moreover, the semantics of verbs is organized in a way that depends on the language concerned without being arbitrary. Morphology and syntax are not direct translations of the archetypal schemas into utterances. Cognitive schematizations are ultra-deep structures.

4.2.2.2 Second Group of Verbs: The Impact of the Subject on the State of Objects and Subjects in Its Environment

Considering the dynamic structure of the meaning of these verbs, we can refer to standard physical models of dynamics (see Kelso [16], who applies this method in cognitive science). The pendulum (like the solar system) is a standard model of

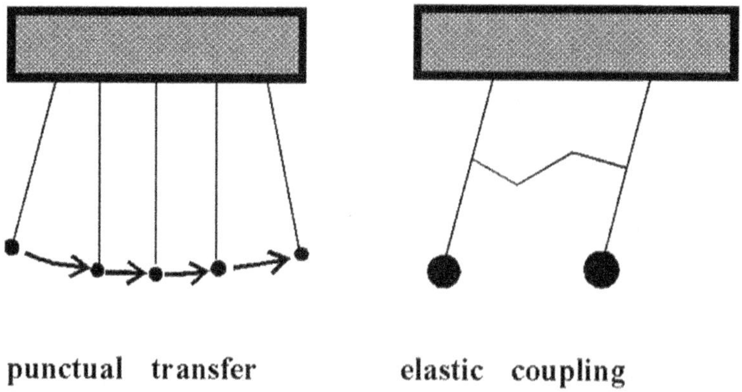

Fig. 4.10 Two types of coupling

dynamics in physics. We will consider the special case of two coupled pendulums and their analogs in the semantics of verbs. Figure 4.10 shows two types of coupled pendulums and, therefore, different types of coupling of two forces.

Whoever pushes a rigid object must give energy corresponding to the friction and the work of displacement; moreover, the rhythm of the application of force may play a role. In the case of the forces applied to a swing, the agent on the swing must be in resonance with the proper rhythm of the swinging object. In the case of a cart's movement, the agent's continuous action is transformed into a cyclic movement of the wheels; the bicycle requires a cyclic movement of the pedal and transforms it over several stages into a straight movement. For semantic schematization, it is a question of finding the qualitative features of the processes present at the motor control level (which is linked to the physical aspect of the process) and on the level of the categorical perception of the process and its verbalization.

This controlled release of energy can be momentary, continuous, or repeated.

(a) Momentary effect:

- to throw, shoot, kick,
- to catch, seize on the fly, collect, stop.

(b) Continuous effect:

- to push, pull, drag, extract,
- to brake, slow down.

Under the configuration aspect, one obtains a dynamic figure corresponding to the emission and capture archetypes. Under the aspect of intentionality, the structure is more complicated. We have to consider transference processes, therefore, the archetype of the gift.[24] Figure 4.11 gives some examples of an (energetic) effect of

[24] Cf. for this archetype and its derivation Wildgen ([47]: 62–82; in English) and Wildgen ([49]: 169–197; in German).

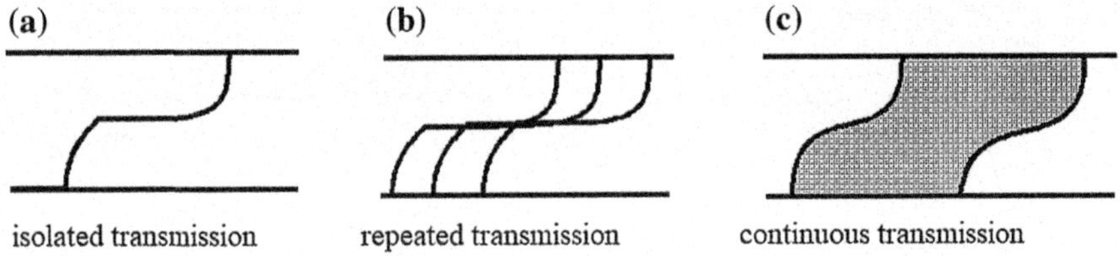

Fig. 4.11 Different transmission configurations

actant one on actant two; this effect can be punctual, iterated (rhythmic), or even of a dense (continuous) nature:

In case (a), it is an isolated energy effect, e.g., "touching sth/sth", the transmission can be carried out with the help of a limb (fingers, lips) or an instrument; in case (b) a repeated energetic effect is necessary, in many cases the rhythm of the application of the force must be adapted to the rhythm of the object/person undergoing the effect. This is the case for "pushing a swing". Finally, in case (c), the continuous application of energy is necessary. For example, "push a cupboard/a car". Finally, the dynamic coupling can link two human subjects (one of whom will be the agent, the other the patient). In this case, the transmitted energy is intentional or performative (communicative).

The second group: (creation, destruction) of Ballmer and Brennenstuhl refers to a sort of homeostasis of the system, controlled by emissions, captures, gains, and losses of order (of information, of negentropy).

4.2.2.3 Third Group of Verbs: Changes/Transitions in the Space of Qualities

The influence of the subject or agent can trigger a change on a qualitative scale; the patient is modified according to one of his qualities. We will give an example by choosing the subgroup of German verbs classified by Ballmer and Brennenstuhl [4] and called "regeneration/disintegration" as an empirical reference. It refers to different quality spaces.

(a) verbiegen (distort, twist)
(b) reinigen (cleanse)

The processes that are classified by verbs refer to a qualitative space with the following states:

(+) straight (–) twisted "verbiegen" (distort).

(+) clean (–) dirty "reinigen" (to clean).

In the control space, we have paths that go from:

(a): (+) → (–): verbiegen (twist)
(b): (–) → (+): reinigen (clean)

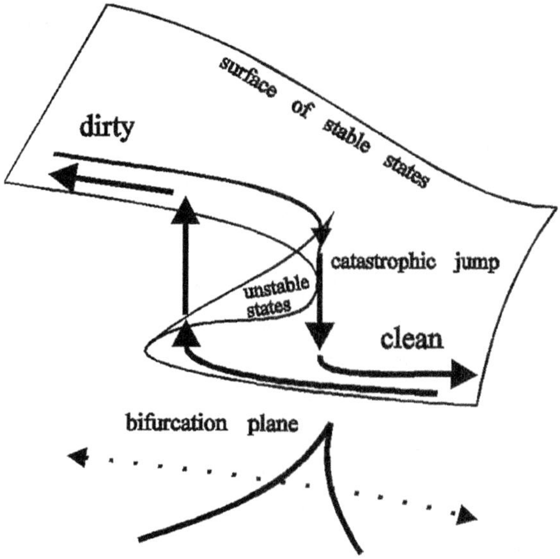

Fig. 4.12 The dynamics of disaggregation and regeneration (example "cleaning")

Figure 4.12 shows the dynamic modeling.[25]

The first group (impact of a subject on a state) has as its content the coupling of the dynamic system of an agent with another system affecting the state, the energy, and the intention of the actant in the 'object' position. Semantic mapping can exploit these energetic, intentional, and qualitative differences. However, whether specific features are selected for verbalization depends on the inherent needs of linguistic expression in a linguistic community. This situation can be compared to the development of the lexicon of colors, where initially, physiological prefigurations play a major role. However, in the elaboration phase, expressive need and cultural conditions govern all further differentiations (see Wildgen [47]: 97f. and the contributions in Plümacher and Holz [26]).

[25] See Thom ([33]: 208) where the verb "to wash" is schematized. In this context, an opposition between "clean" and "unclean" is introduced. The dirt is first bound to the unclean object and then removed by a catastrophe of excision; the remaining object is "clean". We can consider our schematization as a local restriction of the archetype of excision in which the actant (the object) changes quality: O(dirty) → O(clean); see Fig. 8.7 in Thom (ibidem). The archetype of "excision" mentioned by Thom is derived from a linear path in the hyperbolic umbilic. In Wildgen ([47] and [49]) an explicit derivation of the four valent archetype of "sending" is given. The derivation of the other archetypes referring to the umbilics was not possible given the state of the art in this period and no actual results are known to the author.

4.3 Sketch of a Morphogenetic Explanation of Human Language

The transformation of Darwinism in the late seventies of the twentieth century led to new theories called EVO-DEVO (ecological evolutionary developmental biology).[26] They point to deep homologies that link biological entities separated by hundreds of millions of years, for instance, fruit flies, mice, and humans. The key to these analogies is ancient genes that have been conserved for millions of years and can control the homology of basic morphological "plans" despite huge morphological differences. Homeotic genes, i.e., genes that control the identity and correct order of bodily segments, constitute the "homeobox", i.e., the set of gene segments shared by the homeotic genes.[27] Many of the homologies mentioned by René Thom in 1972 between patterns in plants and animals on the one side and humans and human behavior on the other can be related to such ancient genetic programs (morphogens, instructed signals). However, experimental research in genetics cannot bridge the gap between unity and diversity in fruit flies or mice and universality/diversity in language and culture.

Nevertheless, we may conjecture that the organization of perceptual organs (in humans, particularly sight and hearing) constitutes a kind of "homeobox" for the unfolding of symbolic forms, specifically human language. A further ancient base could be motion, locomotion, and spatial orientation. These two resources are candidates for universal features genetically coded on a homeobox; indirectly, human behavior and cultures may be controlled by the "old" resources: perception and motion. Moreover, they constitute the conservative basis established millions of years ago that made the evolution of language and other symbolic forms possible. For more details, see Wildgen [54] on the evolution of language and Wildgen [41] on the morphogenesis of symbolic forms.

In the grammar of human languages, invariant form-giving principles constitute a "homeobox" responsible for homologies between human languages. We do not assume the existence of universal grammar (based on innate "ideas") but a gradient in the invention and selection of grammatical forms that leads to the statistical prominence of certain types of grammatical organization (constructions). Such an evolution does not mean grammatical structures whose genesis can take thousands of years are the immediate outcome of specific morphogenetic processes. One must instead assume myriads of communicative acts involving entire populations that create ad hoc lexical and syntactic patterns, select among a given inventory, and change them in detail. The cumulative outcome of these processes selects and optimizes certain organization features in the lexicon and the syntax.[28] The details of such a long-term

[26] This section takes up contents exposed in more details in Wildgen ([41]: chapters 1 and 6).

[27] See Bailleul et al. [3].

[28] Evidence from language change documented by historical and comparative linguistics indicates that stability is maintained against forces of destabilization or new optima are sought. For example, Steels (2014) simulated the language change from Old High German (ca.800–1000 AD) to New High German (ca. 1650–today) using computer agents in a language game. He concludes, "Although

process with many degrees of freedom cannot be reconstructed in linguistics (in the present state). Therefore, the morphogenetic analysis of grammar must take a detour and use the morphogenetic patterns abstracted from nature and seek plausible patterns discovered in languages until now that are possible realizations of these abstract morphogenetic structures. Finding the most basic entities underlying a set of complex structures is similar to the search for basic figures in geometry. The analogy between geometry and psychobiological phenomena was already known to Aristotle (in his treatise on psychology, "De Anima", Aristoteles [1]: § 414b16). The abstract geometrical character of elementary representations is also evident in neurophysiology, and the psychology of vision; cf. for the discussion of this research and its relation to Thomian thought [24]. The analysis of sensory inputs consists of mappings from a three-dimensional input into a precise control of activity in space and time. The mapping must conserve essential topological and dynamic characteristics and can forget metrical details and variations of a type of object or event. Therefore, the problem of structurally stable mapping lies at the heart of every theory of representation and semantics.

4.4 Geometric and Dynamic Morphologies of Visual Signs

The geometric nature of visual signs (painting, sculpture, film, architecture) is immediate.[29] The different visual media are distinguished in the extent and how they represent the dynamics inherent in the objects of meaning, the referents. The painting must project a three-dimensional reality into two dimensions and find indirect ways to convey the third dimension. The dimension of time may be absent, or the painting contains clues that allow the observer to reconstruct the absent dynamic. In the case of sculpture, three-dimensionality is guaranteed (if the sculpture can be viewed from all sides); the observer is not motionless in front of the sculpture as in front of the painting. However, the dynamic inherent in the object generally remains limited. What we have said about the painting is also valid for photography. However, only film (video) introduced explicit dynamics into visual media. In what follows, We shall recall only a few traits reported in Wildgen [58].

The organization of the painted surface is traditionally governed by geometric principles, such as the role of the diagonal, left–right, and top–bottom asymmetry and their symbolic values. In the center, a triangular (or pyramidal) structure dominates many of Leonardo's paintings (see Wildgen [58]: 89–100). In modern art (semi-abstract or abstract), the interplay of curves and curvilinear surfaces becomes the

---

there have been phonetic processes (unrelated to function) that have eroded endings and merged forms, only those solutions that lead to a more optimal system from the viewpoint of semantics, morphosyntax, and phonology have undergone positive selection."(ibidem 348).

[29] Visual semiotics was treated extensively in Wildgen ([58]; in German). Major aspect are taken up in Wildgen ([41]: chapter 4; in English). See also (in French): Wildgen [55, 39, 60, 42].

support for signifiers and signifieds. A curvilinear surface can be understood as the deformation of a surface of the circle or ellipse type (see Leyton [17]).

Finally, some paintings of abstract art, as in the case of Jackson Pollock (1912–1956), see the painting: Number 32 (1950), lack a geometric structure (neither center nor diagonals). Instead, they are the product of an artist's movement, which causes the color to flow over an extended horizontal surface. In the case of Pollock, one can even speak of fractal art (see Wildgen [39] and [58]). The visible dynamic refers to the movements/actions of the artist.

Regarding topology, the naive drawing (of a child, of a drawing amateur) is locally metric. This is even valid for all paintings before the perfection of perspective during the Renaissance (Filippo Brunelleschi and Leon Battista Alberti). However, with the introduction of photography, visual art abandoned this realistic "perfection", and simpler topologies replaced the metric. Apparent contours were even reversed. Eventually, cultivating colored surfaces became the original aim, and "impressionist" effects, networks of contrasts, and other visual effects enlarged the realm of visual content (see Wildgen [58]: chapter 4 for semiotic innovations in art since the 19th century).

The visual dynamics in movies concern, first of all, the signifier, that is to say, what we see is in motion, and we can recognize forces and the effect of forces (the dynamics). The signifieds are equipped with mental kinematics and dynamics. They capture the significance of what we see and shape the cognitive and emotional values induced by the movies (see Wildgen [59]).

4.5 Geometric and Dynamic Morphologies of Musical Signs

We can distinguish three dimensions and their topological and dynamic aspects:

(1) The rhythm of music corresponds to structures of human movement (movement of the legs while walking, running, dancing; movement of the hands in work, in craft performance). Musical beats like the march (2/4) or the waltz (3/4) also reflect the rhythms of locomotion. Another source of rhythm relates to the heart rate and breathing rate. This correspondence has the effect that the musical rhythm can influence the rhythm of locomotion, for example, of a group of musicians or dancers. In other cases, it influences the heart rate, accelerating or slowing down. Other bodily rhythms are influenced by music and have repercussions on the well-being or discomfort of performers or listeners of the music. In all these cases, the rhythm is a signifier of the music, and the bodily and emotional effects determine the signified.

(2) Pitch movements have certain general characteristics: the pitch rises from low to high or vice versa, it remains in a low or high zone, it advances in small steps or jumps, or it wobbles around an average level. These movements are associated with emotions. A piece of music, for example, by Mozart, can evoke

a journey through the emotional landscape with stopping points, transitions, slowing down and speeding up, and a final stop.

The melodic phrase has an intentional character in that it makes the listener feel a theme to which he should pay attention. This quality suggests an analogy with the deictic gestures made with the hand or by the movement of the gaze. The meaning of the musical phrase is a moment of attention that reinforces the perceptual effect. As a result, the musical phrase can be recognized when it reappears during a variation or transposition. This effect allows the organization of more complex structures with a wider temporal scope. For example, the music creates complex paths through counterpoint in Bach's fugues.

From the point of view of differential topology, the temporal organization of music concerns the periodicity of musical events (rhythm, pitch, musical phases). In a popular song, the repetition of the refrain and the repetitive structure of the lines of a song display a simple periodicity. At the same time, a fugue by Bach articulates multiple and parallel periods with its counterpoint technique. Schönberg's atonal compositions replace the structural invariant of tonality with a new periodicity that reduces tonal repetitions and opens up a free type of seriality. See Wildgen ([62]: 113–125).

We did not develop the geometric and dynamic modeling for the visual and musical examples, as this would have required a discussion of the diversity of musical forms in history and a formal technicality beyond that introduced in the previous chapter. To conclude, we want to return to the problem of mathematization in the sciences, focusing on semiotics.

4.6 The Relevance of (Qualitative) Mathematics for Semiotics

If J.S. Mills says: "a great mathematician is not only a mathematician",[30] one can infer that a semiotician, interested in the development of his discipline, should be something more than a semiotician and take an interest in neighboring theoretical disciplines: philosophy and mathematics (or logic). In the case of philosophy, the necessity of an enlarged scope seems obvious, as the examples of Charles Sanders Peirce, Umberto Eco, and others show. As for mathematics, such a larger interest is less obvious. The major currents of semiotics in the 20th century were oriented toward literature (Greimas), history, or the arts (visual and musical; cf. Greimas). Therefore, intellectual areas stood in the intellectual focus where we hardly think of mathematical applications. There are, however, sub-fields, for example, computer art and electronic music, which require mathematical theorizing; this also applies to linguistic engineering. Close to semiotics, philosophy experienced a mathematization in Carnap's logical empiricism and the analytical philosophy that dominated

[30] See the preface by Jean Largeault to the volume "Apologie du logos", Thom [34].

philosophical thought in the United States in the second half of the 20th century and was disseminated in Europe towards the end of the 20th century. Neuro-philosophy has continued this trend. We must therefore ask what role mathematics of the type used by Thom can play in semiotics. The applications of mathematics in the natural sciences have characterized modern science since Galileo and Newton. However, it was not obvious that the human and social sciences could use these means since quantitative measurements and experimentation prove difficult in these disciplines.

The fundamental activity of the scientist consists in finding the relevant categories which allow him to grasp the phenomena of his field of observation. The naive and first method consists in inventing a terminology (a specific lexicon) to talk about these phenomena. Categorization, however, requires an operation of abstraction in which the relevant criteria are distinguished and where the impact of secondary and tertiary criteria is reduced. The concept of "structural stability" (under the impact of a dynamic of fluctuations, changes, and variations) captures this action of abstraction (it is insufficient to arbitrarily neglect some phenomena, as done in Saussurian or Chomskyan linguistics). As the variations often have a stochastic character, the notion of "slaving factors" in Herman Haken captures this moment in a general way, and it includes the morphologies of Thom while placing them in a stochastic context (it thus brings together the morpho-dynamics of Thom and the thermodynamics of Prigogine).

Mathematical forms allow it to go beyond appearances and advocate principles or even deeper laws, which have a more global scope than classifications and descriptions based on disciplinary observations.[31] The conceptual core of the applications of geometry and topology refers to the concept of form (εἶδος, eidos) and morphology (see μορφή, morphé), which deals with forms. In living forms, these are always changing, moving, and in action; we speak of "morphodynamics". Since Galileo's kinematics, the sciences have mastered the kinematic aspects, the evolutions over time, and the dynamic aspects (the effect of forces) responsible for them. However, most logical systems have not taken this turn, and they remain in a static perspective, in a fixed (eternal) world. If we compare these trends in science to those in visual art, we can say that, on the one hand, cinematography has set visual representations in motion. On the other hand, morphodynamics in the human sciences is beginning to set scientific representations in motion. This innovation applies specifically to semiotic phenomena, for instance, linguistic, visual, and musical phenomena.[32]

[31] This strategy refers to Plato and, more recently, to Kant and his concept of a synthetic a priori, therefore of a general structure without experience support but which allows us to deepen our knowledge based on experience. Petitot [23] and [20] elaborated this philosophical aspect of catastrophe theory.

[32] After the book: "Visuelle Semiotik" (visual semiotics) in 2013 and "Semiotik der Musik" (semiotics of music) in 2018, the author published a third volume: "Mythos und Religion. Semiotik des Transzendenten" (Myth and Religion. Semiotics of the Transcendent); see Wildgen [64]. Major contents of this trilogy in German were summarized in the Chaps. 3–5 of Wildgen [41].

References

1. Aristotle: de Anima. In: Shields, C.J. (ed.) Clarendon Press, Oxford U.P., Oxford (2016)
2. Arnold, V.I.: Catastrophe Theory. Springer, Heidelberg (1986)
3. Bailleul, R., Manceau, M., Touboul, J.: A "Numerical Evo-Devo" synthesis for the identification of pattern-forming factors. Cells **9**(8) (2020)
4. Ballmer, T.T., Brennenstuhl, W.: Deutsche Verben. Eine sprachanalytische Untersuchung des Deutschen Wortschatzes, Narr, Tübingen (1986)
5. Brandt, P.A.: La charpente modale du sens. Pour une sémio-linguistique morphogénétique et dynamique. Aarhus U.P., Aarhus (central chapters of a »thèse d'état » 1987, Sorbonne, Paris) (1992)
6. Brandt, P.A.: Spaces, Domains, and Meaning. Essays in Cognitive Semiotics, Lang, Bern (2004)
7. Bruter, C.P.: Topologie et perception. Bases philosophiques et mathématiques. Maloine, Paris (1974)
8. Chomsky, N.: Syntactic Structures. Mouton, Den Haag (1957)
9. Chomsky, N.: Aspects of the Theory of Syntax. MIT Press, Cambridge, Mass (1965)
10. Chomsky, N.: The Logical Structure of Linguistic Theory, manuscript, electronic copy, http://alpha-leonis.lids.mit.edu/chomsky (partial publication, Plenum Press, New York, 1975) (1955)
11. Gilmore, R.: Catastrophe Theory for Scientists and Engineers. Wiley, New York (1980)
12. Greimas, A.J., Courtés J. (éds.): Sémiotique: dictionnaire raisonné de la théorie du langage, T. 2. Hachette, Paris (1986)
13. Guckenheimer, H., Holmes, P.: Nonlinear Oscillations, Dynamical Systems, and Bifurcations of Vector Fields. Springer, Berlin (1983)
14. Haken, H.: Synergetics. An Introduction (third edition). Springer, Berlin (1983)
15. Haken, H.: Brain Dynamics. An Introduction to Models and Simulations. Springer, Berlin (2008)
16. Kelso J.A.S.: Dynamic Patterns. The Self-Organization of Brain and Behavior. Bradford, Cambridge (Mass.) (1997)
17. Leyton, M.: The mathematical structure of pain. In: Wildgen, W., van Heusden, B. (éds.) 2009 Meta-Representation, Self-organization and Art, Lang, Bern, pp. 83–114 (2009)
18. Mandelbrot, B.B.: The Fractal Geometry of Nature. Freeman, New York (1979)
19. Petitot, J. (ed.): Logos et Théorie des Catastrophes. Patiño, Geneva (1988)
20. Petitot, J.: Physique du Sens. Presses du CNRS, Paris (1992)
21. Petitot, J.: Topologie du carré sémiotique. In: Études littéraires, Université de Laval, Québec (1977)
22. Petitot, J.: Pour un schématisme de la structure: De quelques implications épistémologiques de la théorie des catastrophes. Thesis, École des Hautes Études en Sciences Sociales, Paris (manuscript) (1982)
23. Petitot, J.: Morphogenèse du sens. Pour un schématisme de la structure, Presses universitaires de France, Paris (central chapters of Petitot 1982) (1985)
24. Petitot, J.: Neurogéométrie de la vision. Modèles mathématiques et physiques des architectures fonctionnelles, Les Éditions de l'École Polytechnique, Paris. English translation. Elements of Neurogeometry: Functional Architectures of Vision (Lecture Notes in Morphogenesis). Springer, Cham (2008)
25. Petitot, J.: Les premiers textes de René Thom sur la morphogenèse et la linguistique: 1966–1970. HALopen science, hal-01265180 (2015). https://hal.science/hal-01265180v2
26. Plümacher, M., Peter, H. (eds.): Speaking of Colors and Odors. Benjamins, Amsterdam (2007)
27. Poston, T., Stewart, I.: Catastrophe Theory and its Applications. Pitman, Boston (1978)
28. Prigogine, I., Stengers, I.: La nouvelle alliance: métamorphose de la science (2nd edition, Gallimard, Paris, 2005) (1979)
29. Slodowy, P.: Platonic solids, kleinian singularities, elementary catastrophes, and lie groups. In: Petitot, J. (ed.) A partir de l'œuvre de René Thom, Patiño, Geneva, pp. 73–98 (1988)
30. Sussmann, H.J., Zahler, R.S.: Catastrophe theory as applied to the social and biological sciences: a critique. Synthese **37**, 117–216 (1978)

31. Tesnière, L.: Eléments de syntaxe structurale. Klincksieck, Paris (first edition 1959) (1988)
32. Thom, R.: La linguistique, discipline morphologique exemplaire. Critique **30**(1), 235–245 (1974)
33. Thom, R.: Esquisse d'une Sémiophysique. InterEditions, Paris (1988)
34. Thom, R.: Apologie du logos. Hachette, Paris (1990)
35. Thom, R.: Topologie et linguistique. In: Haefliger, A., Narasinkan, R. (eds.) Essays in Topology and Related Topics, pp. 226–248. Springer, Berlin (1970)
36. Thom, R.: Stabilité structurelle et morphogenèse. Interéditions, Paris (second edition corrected and augmented, 1977) (1972)
37. Thom, R.: Modèles mathématiques de la morphogenèse: recueil de textes sur la théorie des catastrophes et ses applications. Union Générale d'Éditions, 10/18, Paris (1974b)
38. Waddington, C.H.: The character of biological form. In: Whyte, L.L. (éd.) Aspects of Form. A Symposium on Form in Nature and Art. Second edition 1968 (the first edition had been published in 1951) (1968)
39. Wildgen, W.: Geometry and dynamics in the art of Leonardo da Vinci. Cogn. Semiot. **5**, 1–30 (2010)
40. Wildgen, W.: Giordano Bruno. LIT-Verlag, Berlin, Neun Studien und Dialoge zu einem extremen Denker (2011)
41. Wildgen, W.: Morphogenesis of Symbolic Forms: Meaning in Music, Art, Religion, and Language, Cham (C.H.): Springer Nature. Lecture Notes in Morphogenesis, Series (2023)
42. Wildgen, W.: L'autre de la sémiotique du langage: Les signifiés visuels et musicaux et leur caractère morpho-dynamique. In: Biglari, A., Roelens, N. (eds.) La sémiotique et son autre, pp. 451–468. Éditions Kimé, Paris (2019)
43. Wildgen, W.: La linguistique est-elle « rigoureusement inutile »? Réflexions sur un dicton de René Thom. In: Papadopoulos, A. (ed.) René Thom, pp. 367–377. Presses du CNRS, Paris, Portrait mathématique et philosophique (2018)
44. Wildgen, W.: Semantic ambiguity in relation to perceptual multistability. In: Stadler, M., Kruse, P. (eds.) Ambiguity in Mind and Nature, pp. 221–240. Springer, Berlin (1995)
45. Wildgen, W.: Verständigungsdynamik: Bausteine für ein dynamisches Sprachmodell. Thesis of Habilitation, University of Regensburg (557 p., électronic copy) (1979)
46. Wildgen, W.: Archetypical dynamics in word semantics: an application of catastrophe theory. In: Eikmeyer, H.-J., Hannes, R. (eds.) Words, Worlds, and Contexts. New Approaches to Word Semantics, pp. 234–296. de Gruyter, Berlin (1981)
47. Wildgen, W.: Catastrophe Theoretical Semantics. An Elaboration and Application of René Thom's Theory, Benjamins, Amsterdam (1982)
48. Wildgen, W.: Goethe als Wegbereiter einer universalen Morphologie (unter besonderer Berücksichtigung der Sprachform), pp. 235–277. Jahresbericht des Präsidenten 1982, University of Bayreuth (1983)
49. Wildgen, W.: Archetypensemantik. Grundlagen für eine dynamische Semantik auf der Basis der Katastrophentheorie, Narr, Tübingen (partial publication oft he thesis of 1979) (1985)
50. Wildgen, W.: Basic Principles of Self-organization in Language. In: Herman, H., Michael, S. (eds.) Synergetics of Cognition, pp. 415–426. Springer, Berlin (1990)
51. Wildgen, W.: Process, Image, and Meaning. A Realistic Model of the Meanings of Sentences and Narrative Texts. Series: Pragmatics and Beyond, New Series, No. 31, Benjamins, Amsterdam (1994)
52. Wildgen, W.: Das kosmische Gedächtnis. Kosmologie, Semiotik und Gedächtnistheorie im Werke von Giordano Bruno (1548–1600), Lang, Frankfurt (1998)
53. Wildgen, W.: De la grammaire au discours. Une approche morphodynamique, Lang, Bern (1999)
54. Wildgen, W.: The Evolution of Human Languages. Scenarios, Principles, and Cultural Dynamics, Benjamins, Amsterdam (2004a)
55. Wildgen, W.: Éléments narratifs et argumentatifs dans l'articulation de l'espace pictural. Transformations de l'Ultime Cène du XIIe au XXe siècle. In: Stefania C. (éd.) Espaces perçus, territoires imagés en art, pp. 77–97. L'Harmattan, Paris (2004b)

56. Wildgen, W.: Kognitive Grammatik. Klassische Paradigmen und neue Perspektiven, de Gruyter, Berlin (2008)
57. Wildgen, W.: Die Sprachwissenschaft des 20. Jh.s: Versuch einer Bilanz, de Gruyter, Berlin (2010a)
58. Wildgen, W.: Visuelle Semiotik. Die Entfaltung des Sichtbaren. Vom Höhlenbild bis zur modernen Stadt, transcript, Bielefeld (2013)
59. Wildgen, W.: "Movie Physics" or dynamic patterns as the skeleton of movies. In: Wildfeuer, J., John A.B. (éds.) Film Text Analysis. New Perspectives on the Analysis of Filmic Meaning (Routledge Advances in Film Studies), pp. 66–93. Routledge, New York/London (2016)
60. Wildgen, W.: Une sémiotique de l'image et de la musique au-delà de la « sémiotique logocentrique » de Greimas. Paper read at the Congress: „Greimas aujourd'hui: L'avenir de la structure", Paris (2017a)
61. Wildgen, W.: En cas de catastrophe. Les systèmes casuels et la dynamique qualitative, Contribution au « Colloque Petitot », Paris-Nanterre, 29 mai 2015; published in June 2017. Estudos Semióticos **13**(1), 1–15. https://www.revistas.usp.br/esse/article/download/138414/133874 (2017b)
62. Wildgen, W.: Musiksemiotik: musikalische Zeichen, Kognition und Sprache. Königshausen & Neumann, Würzburg (2018a)
63. Wildgen, W.: Structures, archetypes, and symbolic forms. applied mathematics in linguistics and semiotics. In: Alberto P., Silvano Z.C. (éds.) 2020 Structures Mères, pp. 165–185. Semantics, Mathematics and Cognitive Sciences, Springer, Cham (2020)
64. Wildgen, W.: Mythos und Religion. Semiotik der Transzendenz, Königshausen & Neumann, Würzburg (2021)
65. Zeeman, C.: Catastrophe Theory: Selected Papers 1972–1977. Addison-Wesley, Cambridge (Mass.) 1977

Part II
René Thom Between Geography and Urbanism

Chapter 5
What Can Catastrophe Theory Contribute to the Study of Cities? Two Examples and a Question

Olivier Bonin

5.1 Introduction

Geography is a science concerned with forms: the forms of the earth and the forms of its settlements, if we understand the term form as the way of filling space. It is therefore legitimate to ask how a resolutely structuralist approach such as that of René Thom's catastrophe theory, which places form as the central object and claims to be relatively universal, can shed light on geographical issues. The concept of form in catastrophe theory is not entirely identical to that of form in geography, and these two notions will interact in an original way in the models proposed here.

Although I am not sure that I have identified all the texts on the subject, it has to be said that few articles and approaches in geography are directly based on catastrophe theory—well under a hundred to my knowledge. Wilson [13] and Rosser [12] offer summaries of these applications to the regional sciences, geography's close cousins. In general, the aim is to model the growth of cities, an attempt which we will take up again in the second application proposed here. Among these works are those by Mees [7], who attributes urbanisation trends to transport costs, Papageorgiou [8], who introduces agglomeration economies as an explanatory factor, and Wilson himself, who in his book introduces a fold catastrophe to link agglomeration economies to city size and congestion. Finally, Cassetti [2] analyses the transitions between the world of the seventeenth century and that of the twentieth century using a cusp.

There is, however, a resolutely structuralist current of geography, that of "structural geography", initiated by Ritchot in the late 1970s [11] and developed by a number of his students and close collaborators. Although very close to catastrophe theory, it is not a direct application of it, and has developed in various directions: semiotics, economics, architecture [3].

O. Bonin (✉)
Univ Gustave Eiffel, Ecole Des Ponts, LVMT, Champs-Sur-Marne, France
e-mail: olivier.bonin@univ-eiffel.fr

© The Author(s), under exclusive license to Springer Nature Switzerland AG 2024

I. Marcos and C. Morier (eds.), *The Relevance of René Thom*, Lecture Notes in Morphogenesis, https://doi.org/10.1007/978-3-031-54982-3_5

Several explanations can be put forward to explain the relative lack of interest among geographers in Thom's catastrophe theory. Firstly, structuralism may seem outdated today, and is strongly decried by part of the scientific community. Lévy [6], for example, believes that "neo-structuralism is impoverishing" when we seek to understand the social, which is a problem if we are interested in human geography. René Thom's theory of catastrophes can seem either too complicated or too simple: too complicated if you want to understand the mathematics on which it is based, or too simple if you are content to identify the few emblematic forms of this theory on the basis of simple intuitions. Finally, paradoxically, the fact that it is qualitative can be seen as a flaw: it is impossible to calculate quantities, to study changes, or even to estimate and calibrate any model. Zeeman's enthusiasm for disseminating Thom's work and multiplying the fields of application without always taking all the necessary methodological precautions provoked criticism from eminent mathematicians such as Smale. He believed that a model should not be the first to be applied, but should be derived from the problem being studied. The immense enthusiasm for catastrophe theory has given way to a certain scepticism, which has probably prevented a methodical search for applications in certain disciplines such as geography.

In this article, I propose two applications, and a research question, on issues of urban geography. The first is a necessarily partial and reductive re-reading of the founding ideas of structural geography within the framework of catastrophe theory. This application considers form as a form that unfolds in space, i.e., a form that is not purely abstract. The substratum of this form is geographical space in the usual sense of the term. The second is a project carried out with two engineering students from the Ecole des Ponts in an attempt to understand the differentiation of cities into different archetypes; the form will be more abstract, since we are looking for categories, and the spatiality of these cities will not be taken into account in this model. This work, which is still in its early stages, is more in the tradition of previous work by geographers on catastrophe theory. Finally, in the light of these two models, I propose a research question for further study of catastrophe theory.

5.2 What Can Catastrophe Theory Contribute to Geography?

Before beginning to describe the two applications of catastrophe theory that I am proposing here, I think it would be useful to consider what is expected of such modelling. It is certainly not a question of revealing forms, because in geography forms are visible, apparent and almost too numerous. Nor is it about entering into an explanatory logic, or at least not a strictly causal one, because the limits of environmental determinism or utilitarian approaches to the city are all too obvious: is the city a political project? A marketplace? A place of conflict? Is it determined by religious or symbolic functions? Yes, probably a bit of all of these, but they are only different ways of looking at the urban phenomenon, and one way alone does not capture the whole picture. Finally, it is not a question of being able to calculate quantities (populations, distances, house prices, salaries, etc.) since Thom's models are qualitative and not quantitative.

One of the reasons for using catastrophe theory would be to see it as an extension of game theory. In a zero-sum game, a situation often arises where one player maximises his gain while the other minimises his loss: this is a saddle point, visualised as a regular surface in space. Catastrophe theory shows that this configuration corresponds to a cusp, a more topologically complex surface where different equilibria are possible, equilibria that depend on the previous states of the system. Applied to a geographical problem, catastrophe theory would enable us to understand why certain forms of organisation of the substrate, and only these, can appear and be stable, beyond the obvious forms, but also why there are more stable forms than proposed by an approach based on utility-maximising equilibrium, as in economics. Note that geography and catastrophe theory use the word form in similar but distinct senses, and above all in relation to different spaces. In both cases, form is understood as a synonym for structure, i.e., the way in which the elements of a system are linked and articulated, as in the concept of "musical form". However, for geographers, form—or more precisely, urban form, since the applications considered here concern cities—refers to the spatial arrangement of a city's constituent elements, both tangible and intangible: it extends into the everyday space around us. Catastrophe theory forms are generally abstract, conceived in a structuralist tradition: they are latent theoretical models, and therefore unobservable by definition; these forms are not inscribed in geographical space.

A second reason for using catastrophe theory is to check that a set of dynamics, identified elsewhere, is indeed likely to generate variations in shape that are qualitatively in line with those observed, either directly if they are geographical shapes, or through existing phase transitions, i.e., the possibilities of moving between the different stable regions of the catastrophic ensembles. Without being formally a proof, it is a set of clues that gives weight to the formal framework proposed for analysing the evolution of the shape. This seems to be the most relevant approach here.

5.3 The Structural Path of Human Settlement Through the Prism of Catastrophe Theory

Ritchot, a Quebec geographer, invented a "structural" geography, an original approach, initially in geomorphology [10], transposed, by analogy, to urban geography from 1977, summarised in 1992, then in a book co-authored with Desmarais and Ritchot [3]. He presented his work to Thom at a seminar in 1989, and Desmarais wrote his thesis under the supervision of Ritchot and Petitot, the latter himself being a student of Thom's and a connoisseur of his work, which he applied to semiotics (1977) [9]. However, as far as I know, Ritchot's initial ideas were never presented within the strict framework of catastrophe theory, but developed independently. In fact, in its later developments, structural geography was enriched by Greimas's semiotic theory, Girard's theory of the scapegoat, and other theoretical frameworks.

I therefore propose here a small model more strictly related to the theory of catastrophes, the scope of which will only be academic, based on three ideas that seem to me to be fundamental in Ritchot's work.

The first idea corresponds to a concept, the "vacuum", which is central but difficult to define. The vacuum is an abstract object that breaks down the homogeneity of a geographical substratum, at least from the point of view of how it is perceived by populations. A forbidden place, the vacuum is invested with anthropological values, and is both repulsive, because settlements are forbidden there, and attractive, because people tend to congregate in its vicinity.

The second idea is that of the political control of mobility, mobility being understood as residential mobility. One part of the population manages, directly or indirectly, to gain control over residential mobility, using economic, political, legislative or other tools, while the rest of the population is subject to this political control, which reduces the space in which it can reside through various mechanisms, including property control. Ritchot introduces the terms "endoregulation" and "exoregulation" of residential trajectories, depending on whether or not households and individuals have control over their mobility.

Finally, the third idea is partly linked to the attractive nature of the vacuum. Residential trajectories are not independent: they can be attracted and polarised, with concentration effects around centres, or dispersed and diffused, so as to cover the whole area.

These three concepts are basically quite simple, even if the idea of a vacuum seems to be more abstract, almost metaphysical, than the other two, which are quite concrete and empirically verifiable. By combining them, we can already obtain archetypes of urban settlements. Focalising endoregulation corresponds to a zone of assembly, diffusing endoregulation to a zone of escape, focalising exoregulation to a zone of concentration, and diffusing exoregulation to a zone of dispersion. The types of human settlement result from the neighbourhoods or archetypes of these four zones, giving rise to many intermediate forms. Beautiful neighbourhoods and historic centres are typical of congregation, holiday resorts or gilded suburbs of escape, suburbs of concentration, and finally the countryside of dispersion, and the 'city' is the saddle point between congregation and concentration [11]. This is a rich analysis, but in the end, it takes the approach of game theory with a simple topology: the most complex form that it is possible to obtain is the saddle point, at the junction of a ridge line and a thalweg line of the morphology thus created. The vacuum has not really been used, or at least not in the same way as the other concepts.

I therefore propose to use these three concepts to construct a model for the differentiation of urban positions, as encountered in a city or, more generally, in a human settlement, with the help of catastrophe theory.

To adopt a terminology typical of catastrophe theory, I consider that the urban form is deployed in an abstract space and that the vacuum constitutes the initial singularity of this deployment. The unfolding variables, in this framework, will be axes defined by the two oppositions endoregulation—exoregulation and focalisation—diffusion. The substrate variable is a positional value: it is a value in the sense of an evaluation of a situation, in this case the occupation of the associated spatial position. It is a variable that is by nature localised in geographical space. In catastrophe theory, the elementary catastrophe with one output variable, the positional value, and two input variables, which are the two axes mentioned above, is the cusp, a figure that is no doubt fairly familiar even to those unfamiliar with Thom's work (Fig. 5.1). Let us briefly recall that in catastrophe theory we have an external space, or substrate, which acts as a control space, and a potentially very general space of latent variables (a differentiable manifold), in which dynamics unfold. Catastrophes correspond to bifurcations towards the different stable states of the system thus constituted, which are assumed to be instantaneously obtained.

Thom's catastrophe theory and his classification theorem tell us that, under reasonable conditions, our problem is similar (homeomorphic) to that of studying the function

$$V(x) = x^4 + ax^2 + bx$$

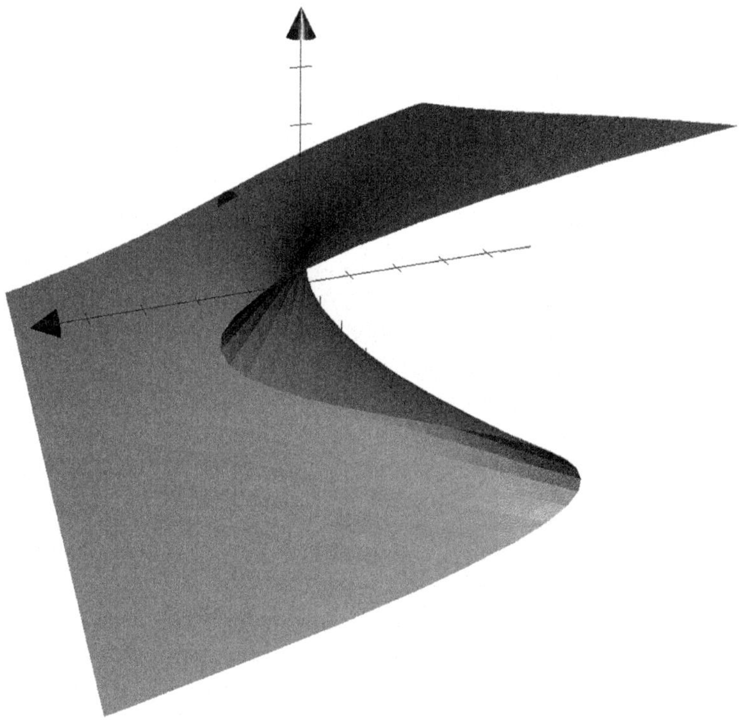

Fig. 5.1 Surface of a cusp catastrophe

with V the value of the system potential and (a, b) the system parameters. The variable x corresponds to the external space, or substrate, in this case the positional value, and the parameters a and b correspond to the latent space, a differentiable manifold in which the dynamics between endoregulation and exoregulation, as well as between focalisation and diffusion, are at work. The strength of Thom's approach is that it would be impossible to explicit the differentiable manifold, let alone the dynamics involved. However, the system has the same properties as that given by the potential $V(x)$, which is quite simple, and the study of its singularities poses no real difficulty.

The equation of state is $\frac{\partial V}{\partial x} = 0$ and the singular points are defined by $\frac{\partial^2 V}{\partial x^2} = 0$, giving as the catastrophic set of the system $a = -6x^2$ and $b = 8x^3$. There is therefore a single stable state outside S, which forms a cusp, and a local maximum and two minima inside S. Outside S is the classic game-theoretic stable fixed point. Within S, there is an unstable fixed point, which is a relative maximum, and two stable fixed points, which are the two relative minima. The transition from one equilibrium to another clearly depends on a change in the variable x, which for us is the positional value.

Catastrophe theory requires us to make a choice and introduce a dissymmetry: a and b are not interchangeable in the potential function, which is visible in the shape of the catastrophic set S, whereas endoregulation—exoregulation and focusing-diffusion are combined symmetrically in structural geography analyses. We therefore need to decide whether a corresponds to the endoregulation—exoregulation or focalisation-diffusion axis. It seems reasonable to postulate that the endoregulation—exoregulation axis plays this role, since it is the primary concept of the structural path of human settlements.

What can we learn from this quickly sketched-out model? First of all, the combinations proposed by Ritchot by crossing the two dimensions are not all stable, and above all this stability depends on the intensity of the positional value. The maximum stable positional value, according to the model, corresponds to situations of endorsement: this is the beautiful neighbourhood or the historic city centre (the "sumptuary massif" in Ritchot's vocabulary), which nothing can shake. Other combinations of factors are also relatively stable, but of lesser positional value: prestige neighbourhoods, city centres, suburban housing estates. In situations of greater exoregulation, we enter the catastrophic ensemble, where there is an unstable maximum and stable minima. The stable minima are easy to identify: the countryside on the one hand, and the suburb, working-class neighbourhood or large housing estate on the other. These are positions where the positional value is low, and there is no internal dynamic in the system we have just posited that would allow us to extract ourselves from these basins without regaining control of mobility, i.e., leaving the catastrophic ensemble. The local, unstable maximum corresponds to the village, whose positional value is resilient, but which can shift at any time to the countryside, during an exodus, or to the peri-urban area when incorporated into an existing urban structure.

The model constructed here is a local model, which describes possible changes in positional value locally. It does not provide a complete description of the spatial organisation of a city, because the spatial gradients of the latent variables (the endorsement or focusing dynamics) are not explicitly known. On the other hand, it does shed light on areas of stability and potential shifts, as well as on the neighbourhoods likely to be observed.

The contribution of catastrophe theory used in this way in this model may seem modest, but it nevertheless makes some small contributions to Ritchot's geography, or confirms the validity of his hypotheses:

- the concept of a vacuum constituting an initial singularity from which human settlements unfold along two axes is effective;
- the elementary catastrophe associated with this model (the cusp) comprises a set of catastrophes that can be interpreted, shedding light on the possible transformations of spaces and their relative stability, providing a break in symmetry in the various possible combinations already analysed by structural geography;
- changes in positional values exogenous to the system, changes in society (e.g., the industrial revolution and parliamentary democracy) do lead to a shift in the system's equilibrium, and can take it into unstable zones.

The shape of the theory of catastrophes here is a cusp: this does not of course mean that the city, even reduced to a positional value, has the shape of a cusp, which would in any case be impossible to conceive of in practice (see Fig. 5.1). The geometrisation of the problem does not concern the shape of the substrate, but rather the shape of the control space that is linked to this substrate. This is a graphic illustration of the difference between geographical shape and catastrophe theory shape noted above.

5.4 An Attempt to Model Urban Development Using Catastrophe Theory

In a second small model, designed with Ennigrou and Zilberberg [5], engineering students at the Ecole des Ponts, we consider urban development by considering cities as elementary objects, reduced to a single quantity, their size, measured by population, population density, or the accessibility of the population to the various urban centres. These cities are qualitatively differentiated. Many city characteristics, such as employment or wages, are correlated with city size, so this approach seems acceptable to us. Here, the city is considered as a human settlement, whose internal structuring is irrelevant to the model. It makes no difference whether the city is monocentric or polycentric, organised into sectors, or even as a finger. Similarly, its possible boundaries play no role in the model.

Indeed, as always, it is difficult to define the precise boundary of a city or urban settlement. Strategies for locating households and activities must be understood from a functional perspective, i.e., by considering agglomerations or open urban areas, while planning decisions are taken within different administrative boundaries, arguing for a more regulatory and closed (but necessarily multi-scale) vision of the city. In this model, the city is an object whose boundary is not clearly defined. Measuring the size of a city is therefore potentially problematic. From the point of view of a digital application, we can consider quantifying the size of the city by measuring population or density, if the definition of a perimeter is relevant, or by measuring the accessibility of the population to the main centres (depending on the monocentric or polycentric configuration of the conurbation), a measure which has the advantage of not being based on a predefined perimeter.

However, considering the size of the city (of the urban settlement) as the variable of interest focuses the model on the question of the local and global resources that will enable a population to remain in place, or to develop. This is the approach that will be favoured, to the detriment of other possible approaches: economic competition between cities, a functional approach to cities, or even urban metabolism. Focusing on the question of the development of cities as a function of resources and their exploitation makes it possible to relate indirectly to the geographical space in which these cities are immersed.

We postulate that the appearance and development of a city in the substratum of geographical space is an unfolding according to axiological qualitative oppositions, oppositions which are necessary to define the space of latent variables. Just as the word "form" is problematic, because it has multiple meanings and is understood in different ways in geography and catastrophe theory, so the adjective "axiological" can be controversial. In catastrophe theory, we define a dynamic on a differentiable manifold; this manifold is provided with a structure. Intuitively, the unfolding variables must be values, in the sense already proposed of evaluating a situation, i.e., perceptions that individuals share with their social groups. This is what enables the manifold to be locally assimilated to a Euclidean space. The term axiological can therefore be understood here in a broader sense than that of philosophers such as Heinrich Rickert: they are not necessarily general values, with a sociological or moral scope.

In the absence of a pre-existing theory, we have adopted the "second way" of catastrophe theory, an abductive approach in which we look for a simple form, an elementary catastrophe, which enables us to obtain the urban-type changes that we can observe. It is important to emphasise this position: we are not starting from a pre-established grid of types or categories of urban settlements, nor from supposed mechanisms, but we are focusing on transitions, with the size of the city as the variable of interest. It may seem contradictory, or even illusory, to study transitions without precisely defining the possible categories, but this is the very principle of complexity approaches, which focus on relationships rather than on the attributes of objects.

A very general theoretical framework is possible, however, by drawing on Dumézil's theory of trifunctionality [4]. The three elementary functions proposed by Dumézil: sovereignty function, war function and production function are relevant to cities, and will allow a more conceptual reading of our model.

As cities are relatively stable, this means that we can rule out a number of catastrophes, such as the elliptic umbilic and the hyperbolic umbilic. Given that we characterise urban development in this model by a single variable, the city's population, the swallowtail is the most interesting candidate model, and it has three parameters, which has led to the selection of three deployment axes. This number of three will also make it possible to attempt to identify Dumézil's three functions.

After analysis, we have identified three main oppositions. We give a relatively pragmatic reading of these oppositions, which we nevertheless attempt to link to more general values using Dumézil's grid, in order to conform to a truly axiological approach:

- it is possible to consider the geographical environment (natural resources, vegetation, relief, hydrography, etc.) as resources to be consumed, from a short-term perspective, by moving away when these resources are exhausted, or on the contrary as a place to be cultivated and developed, by investing it with work, from a long-term perspective; this is the axis of the functions of production and reproduction, with fertility as a value. It is also the classic opposition between nomads and sedentary people; we call this opposition investment—exploitation;
- a city, or a human settlement, can be conceived from the point of view of autonomy, or even autarky, by focusing on the needs of its population, or on the contrary as a member of a globalised world, within a logic of competition and sometimes specialisation; this is the axis of the warlike function, with the value of domination of space. We call this opposition local—global;
- finally, a city may wish to offer a diversity of services, at the risk of housing noisy or polluting functions, or on the contrary focus on the comfort of its inhabitants, rejecting outside its perimeter activities that could potentially degrade the quality of life. This is the function of sovereignty, with the values of autonomy and independence. We call this opposition functional—residential.

This choice is of course debatable, as is the attempt to place it within an axiological framework linked to abstract values. We have deliberately chosen purely qualitative dimensions, since it is the strength of catastrophe theory that it can be based on such differentiable varieties; a classic dynamic system approach would be impossible to establish, as it would not be possible to quantify these axes using continuous variables. Furthermore, we can identify the city archetypes corresponding to the combinations of the axes we are proposing, which is a way of empirically justifying this choice.

Rather than listing the eight possible crossovers, we propose instead to look at the dynamics that correspond to five city archetypes:

- the industrial city is always seen from a global and technical perspective, in order to be able to export its production, and generally from an exploitation perspective; for example, heat-dependent industries (glassmaking, steelmaking, etc.) have historically been located close to coal mines or forests.
- the world city is also a global city, with a focus on knowledge, heritage enhancement, maintenance and development; if it focuses solely on the residential function, with no desire for political domination, it takes the form of a tourist city, or even a museum city;

- the autonomous town, capital, administrative town, necessarily combines a local dimension and a functional aspect, but may be more or less oriented towards exploitation or development;
- the dormitory town focuses on the local and the residential function, with a variable relationship to development; this is a slightly extreme archetype which rarely exists in this form alone.

This first reading shows us that the three proposed axes are operational in discriminating between different types of cities and their evolution over time.

To continue analysing the model posed using catastrophe theory, we now need to study the potential function associated with the swallowtail:

$$U(x) = x^5 + ax^3 + bx^2 + cx$$

where for our application x is the population of the city, and (a, b, c) the control parameters that evolve on the differentiable manifold that we have proposed with the three axes described above. The study of critical points and bifurcation zones, similar to that carried out above, is however a little more technical, as it requires us to calculate roots of polynomials of degrees 3 and 4. The set of bifurcations is $\left(-a; -4x^3 - 2ax; 3x^4 + ax^2\right)$.

To study the set of bifurcations, we need to identify the parameters a, b and c, since they are not interchangeable. We propose to assume that a is the local/global opposition, with for $a < 0$ a global city. We then need to cut the set of bifurcations along the (b, c) plane.

In the local case (Fig. 5.2), a town can stabilise in the form of an autonomous city provided that it offers sufficient services, which is necessarily the case when it has an administrative function such as a county town, and remains attractive to maintain its population. If it specialises in residential functions, it will take the form of a suburban village or small tourist town, thereby losing most of its urban character. The form that combines the local and the farm is unstable, whether it is functional or residential: this is the case of the village, or even the hamlet. It is interesting to note that here we find a result that is very consistent with the previous model.

In the case of a global city, with $a < 0$ (Fig. 5.3), the bifurcations are much more complex, with the appearance of conflict zones. A functional city, focused on exploitation or development, will tend to have the profile of a world city, while a residential city will be a tourist or industrial city. These two zones are stable. The conflict zone is where two minima are possible. It corresponds to cities which, at some stage in their development, may hesitate between several archetypes, corresponding to an industrial city with strong residential development, or an industrial city which is trying to influence the global balance without yet succeeding.

In the conflict zone, there are two distinct zones, with inversions in the hierarchy of potentials corresponding to the two local minima.

Let's take an example to understand what this conflict zone corresponds to. A town in this zone may have developed thanks to the exploitation of a resource, whether local or imported, and may have developed a specific industry dependent on this resource.

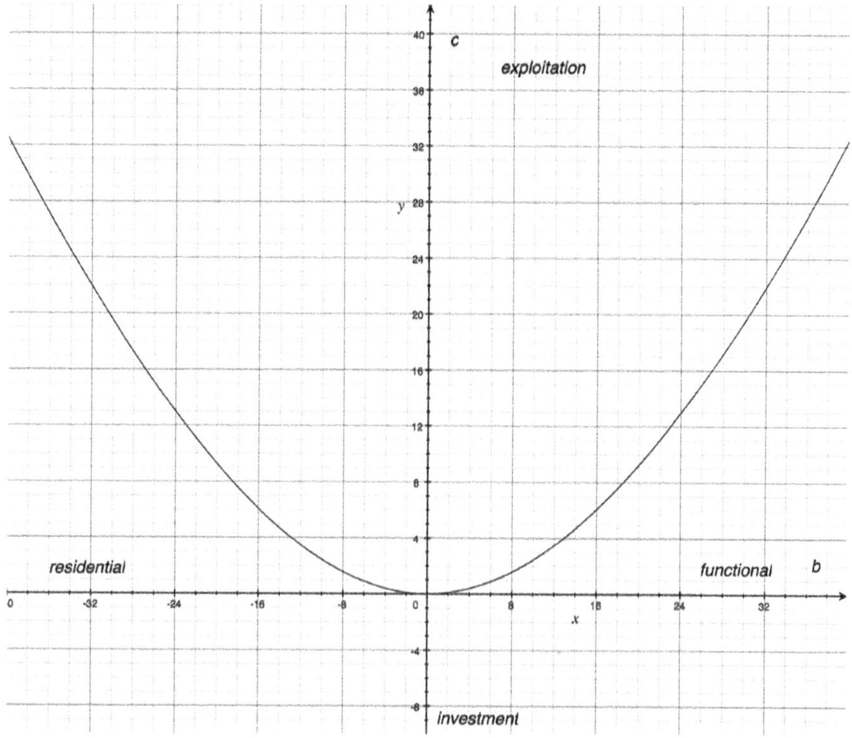

Fig. 5.2 Bifurcation for $a = 10$ (local city)

In conjunction with industrial development, the city acquires new urban or political functions, stabilises itself through its services, trade and the enhancement of its urban capital. It is then an industrial city, in the state of equilibrium of an industrial city, on the left in the graph in Fig. 5.4. Changes in the substrate, i.e., population growth or a strengthening of the connection to the rest of the world through infrastructure, will enable the city to break through the potential barrier and move towards the second equilibrium state, that of a world city, obtained through population growth and corresponding to an increase in trade.

So far, this model has been highly theoretical; now we need to illustrate it with a few examples.

Some towns have gone from the local to the global, like Antibes, thanks to the Sophia-Antipolis technology park and tourism, with the emblematic Marineland. Champs-sur-Marne has developed thanks to the new town of Marne-la-Vallée, but also thanks to Cité Descartes, a major university centre.

The transition from development to exploitation can be seen in Huez, with its mountain tourism (winter and summer), and in Fessenheim, where a nuclear power station has been built.

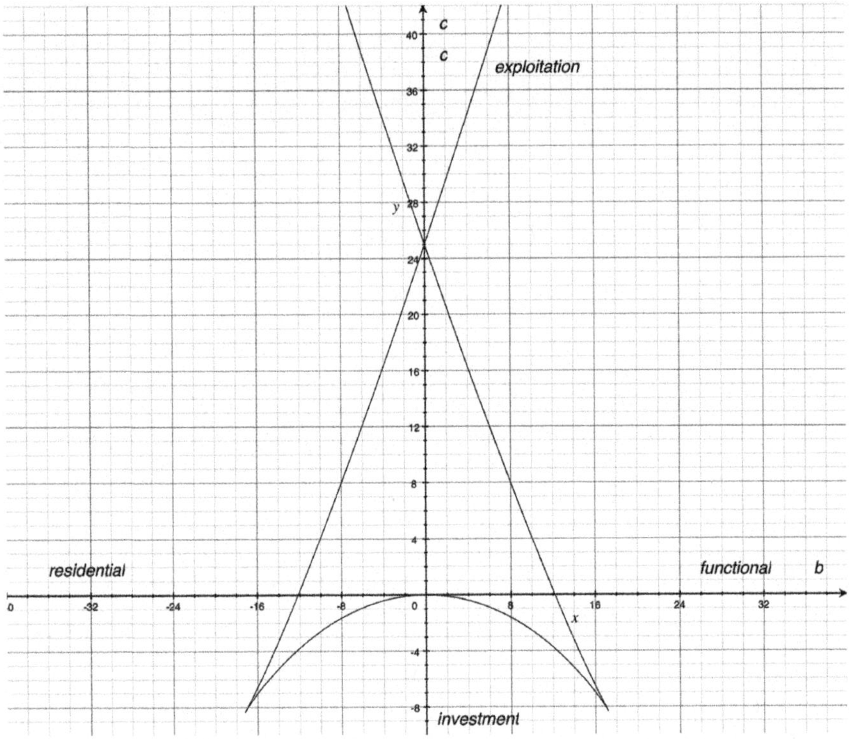

Fig. 5.3 Bifurcation for $a = -10$(global city)

The opposite transition from exploitation to development can be seen in towns that have reacted to deindustrialisation. This is the case for Longwy, a former steel-making town that has been transformed into a dormitory for workers from Luxembourg. Lens is trying to make the most of its mining heritage. The "Louvre Lens" illustrates this desire to make the transition to a value-adding approach.

Finally, some cities have undergone several notable transitions. Lille experienced an initial population peak thanks to the industrial revolution, particularly in textiles and mechanical engineering, followed by several years of stability, then a crisis due to de-industrialisation that led to a shift towards a residential model, before stabilising as a world city at the centre of many flows and exchanges.

The model proposed here is not intended to help predict transitions, by determining the boundary conditions that would trigger them. Purely qualitative, it attempts to explain why certain transitions are more commonly observed than others, and how, among all the possible combinations and transitions between these states, the swallowtail model can significantly reduce complexity.

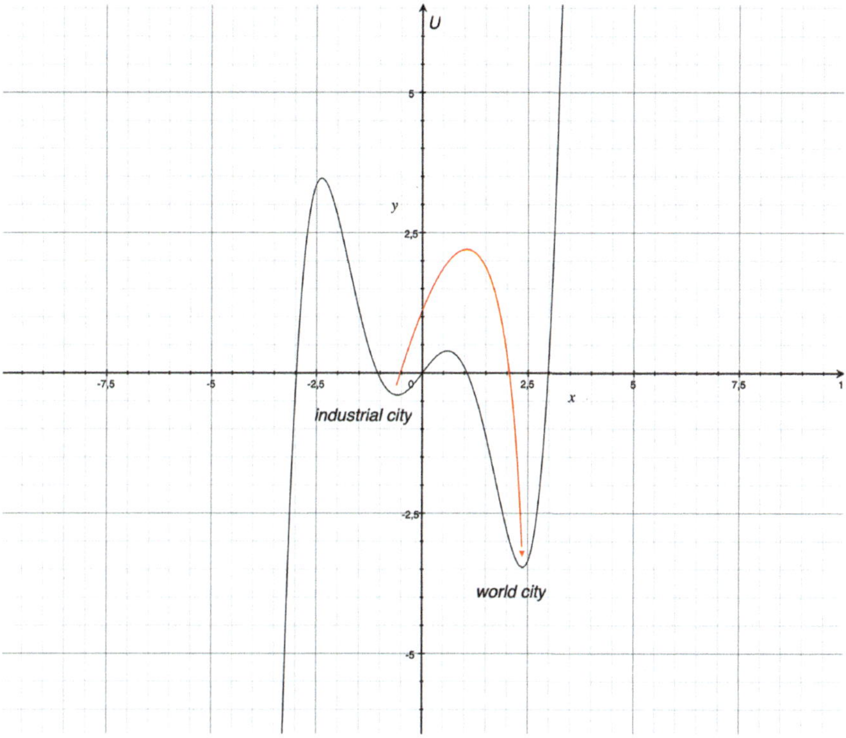

Fig. 5.4 Shape change due to the dynamics of the external parameter ($a = -10$)

5.5 What Does Catastrophe Theory Lack for Geography?

The question may seem presumptuous. The point here is not to identify hypothetical shortcomings in catastrophe theory, something that has already fuelled much debate in the scientific community. It is even less a question of taking sides in this debate. However, the two examples of application raise a question that, as far as I know, has not been much addressed, and which needs to be addressed if we are to continue to develop the applications of Thom's theory to geography.

In catastrophe theory, it is the space of latent variables, controlled by the substrate, which hosts the dynamics, and which selects one of the possible equilibria determined by the form of this differentiable manifold, a form in close correspondence (homeomorphic) with the catastrophes identified by Thom (the fold, the cusp, the swallowtail, etc.) which are the only possible forms, depending on the dimensions of the substrate and the differentiable manifold considered. This approach is very useful for geography, because it allows us to take into account forces that are qualitative in nature, sometimes symbolic, and in any case irreducible to a field of vectors in a vector space.

However, the substrate which, for geography, is generally our space in the usual sense of the term, can be modelled, at least locally, by vector spaces in which the approaches of classical physics (based on positions, trajectories, forces, interactions, etc.) are relevant. This very substrate is subject to forces that have nothing symbolic about them, and which considerably modify the morphogenetic process. These forces are not just gravitation, or the forces of the natural sciences. They are sometimes economic mechanisms; I use the term mechanism deliberately, because the usual mathematics born of mechanics (differential equations, the search for stationary points, etc.) can give a satisfactory account of them.

Let's return for a moment to the first application of the theory of catastrophes proposed here, the one constructed from a few concepts borrowed from Ricthot's human structural geography. This model is purely symbolic: the substrate is a layer of positional value, already unobservable, even if house prices are correlated with this positional value. If, instead of positional values, we wanted to introduce house or land prices, the same construction could be envisaged, with the difference that there is a price dynamic of its own in the substrate, governed by the auction mechanisms of property transactions, and dependent on external macroeconomic variables, such as interest rates and wages, even if this economic dimension is clearly more recent than the symbolic dimension of values. I have in fact proposed such a model with Hubert, but based on another formalism, that of Turing morphogenesis [1].

To apply catastrophe theory to this problem of the evolution of prices coupled to symbolic values, we need to work on the equivalent of the Cartesian product of the vector space in which prices evolve and have their own dynamics by the differentiable manifold in which the forces of endoregulation and exoregulation, as well as focusing and diffusion, are expressed. This mathematical object could be a vector bundle.

I therefore propose as a research question linked to catastrophe theory the investigation of situations where there is a proper dynamic in the substrate, interacting with the proper dynamic in the space of latent variables classically considered. This problem is in fact very general: it opens the door to formal modelling of a physical world (the substrate) and a world of ideas (the latent variables) in interaction. The physical world is governed by a notion of causality, and the world of ideas is governed by a notion of intentionality. These two notions have a different relationship with time, since causality moves down the arrow of time, whereas intentionality moves up.

The theory of catastrophes seems to me to be the ideal starting point for developing a new, more relational theory of complexity, one that coherently articulates the quantitative and the qualitative, the world of forces and the world of forms, from which geography, at the interface between the natural sciences and the human and social sciences, could fully benefit.

References

1. Bonin, O., Hubert, J.-P.: Modélisation morphogénétique de moyen terme des villes: une schématisation du modèle théorique de Ritchot et Desmarais dans le cadre du modèle standard de l'économie urbaine. Revue d'Économie Régionale et Urbaine **2014**, 3 (2014)
2. Cassetti, E.: Testing catastrophe hypotheses. Socio-spatial. Dynamics **2**(2), 65–80 (1991)
3. Desmarais, G., Ritchot, G.: La géographie structurale. L'Harmattan, Paris, France (2000)
4. Dumézil, G.: Jupiter Mars Quirinus–Essai sur la conception indo-européenne de la société et sur les origines de Rome. Gallimard (1941)
5. Ennigrou, M.A., Zilberberg, D.: Projet Initiation à la recherche: Modéliser l'étalement urbain avec la théorie des catastrophes. Projet d'initiation à la recherche de l'Ecole des Ponts (2020)
6. Lévy, J.: Contre la phonie du CACOS, contre le refus d'expliquer le social par le social. EspacesTemps.net [En ligne], Mis en ligne le 9 juillet 2014, consulté le 09.07.2014. https://www.espacestemps.net/articles/contre-la-phonie-du-cacos/ (2014)
7. Mees, A.I.: The revival of cities in medieval Europe. An application of catastrophe theory. Régional Sci. Urban Econ. **5**, 403–425 (1975)
8. Papageorgiou, G.J.: On sudden urban growth. Environ. Plan A **12**, 1035–1050 (1980)
9. Petitot, J.: Topologie du carré sémiotique. Études littéraires **10**(3), 347–426 (1977)
10. Ritchot, G.: Essais de géomorphologie structurale. Les presses de l'Université Laval, 1975, Québec, Canada (1975)
11. Ritchot, G.: La valorisation économique de l'espace géographique. Cahiers de géographie du Québec **36**, 175–214 (1992)
12. Rosser, J.B.: From Catastrophe to Chaos: A General Theory of Economie Discontinuities. Kluwer, Boston (1991)
13. Wilson, A.G.: Catastrophe Theory and Bifurcation: Application to Urban and Regional System. Croom Helm, London (1981)

Chapter 6
Structural, Semiotic and Morphodynamic Geography, Thirty Years On

Jean-Paul Hubert

In March 1989, the semiotics seminar run by Jean Petitot and Jean-Claude Coquet at EHESS, in Paris, successively hosted presentations by Gilles Ritchot, Professor of geography at Québec's Université Laval, and by Gaëtan Desmarais, a geographer and semiotician from Montréal, Ph.D. candidate at EHESS and the kingpin of the event. Then, in May 1989, G. Ritchot presented elements of his theory of structural geomorphology at Bures sur Yvette in the presence of René Thom. The goal of connecting three theoretical ensembles—in human and physical geography, semiotics and mathematics—had been in preparation for several years, and at this point took on new substance and impetus. The undertaking aroused interest, indeed enthusiastic support,[1] and gave rise to publications and PhD memoirs of great diversity in their orientations and disciplines other than geography.[2] Two special issues

[1] Mine, after I attended the presentation by G. Desmarais in 1989. In it, I discovered a way of tackling the discontinuity of geographical space that totally transformed my research perspectives. J. Petitot introduced me to G. Ritchot and G. Desmarais, and the latter was involved in the supervision of my thesis, a role he tackled in the American way with a great deal of commitment.

[2] Especially PhD memoirs on the morphogenesis of geographical spaces (cities, regions or neighbourhoods) with a strong historical component and a semiotic orientation, like those of G. Desmarais in 1991 (Univ. Montréal) and 1993 (EHESS), and of I. Marcos in 1996 (Univ. Aarhus); or oriented towards anthropology: M. Savard 1992 (Univ. Laval) or urban and regional planning: Y. Melançon 1997 (Univ. Laval), S. Gagnon 2001 (Univ. Laval). One thesis dealt with the geometric analysis of cadastral shapes, M. [10] (Univ. Laval),others had a theoretical orientation, examining the theory of urban form from the global perspective of the human sciences: G. Mercier in 1987 (Univ. Laval) with its focus on political economy, J-P Hubert in 1992 (Univ. Paris 1), focused on epistemology, Th. Rebour in 1996 (Univ. Paris 1), on the economy. More recently, R. Guertin defended a PhD in urban planning in 2006 (Univ. Montréal) and L. Ampleman used structural geography to cross geopolitics and transport studies in 2014 (Univ. Jagellon). This last study brings for the first time a full presentation in English of the theory of urban form (2021) [1].

J.-P. Hubert (✉)
Université Gustave Eiffel, IFSTTAR-AME-DEST, Champs-Sur-Marne, France
e-mail: jean-paul.hubert@univ-eiffel.fr

© The Author(s), under exclusive license to Springer Nature Switzerland AG 2024

I. Marcos and C. Morier (eds.), *The Relevance of René Thom*, Lecture Notes in Morphogenesis, https://doi.org/10.1007/978-3-031-54982-3_6

of *Cahiers de géographie du Québec* punctuated this ascending phase, in 1992 (Vol. 36, No. 98) [26] and in 1998 (Vol. 42, No. 117) [7].[3] In 1999, G. Ritchot for the first time expanded the scale of investigation to a large region, with a morphogenesis of Québec as a "form of settlement". The high point seems to have been reached in 2000 with L'Harmattan's publication of the book "Structural Geography" by G. Desmarais and G. Ritchot. However, hopes and numbers of productions subsequently fell, despite the 2011 publication—again by L'Harmattan—of a major new work by G. Ritchot encompassing thousands of years of the West: "Morphogenesis of Rome". Still innovative but isolated and ill, he brought out a final book in 2018: "Champlain's Legacy". He died on January 27, 2021 at the age of 85.

All in all, the trajectory of structural geography is consistent with that of Catastrophe Theory and structuralism in the human sciences. Its model-based and mathematical approach, reflecting its connection with R. Thom's school, was captured in the paradigm of complexity, as was already apparent in the special issue of *Cahiers de géographie du Québec* in 1998. In parallel, ideas about the autonomy of forms as theoretical objects (with geometric properties that have a concrete effect on geographical space by overdetermining the mechanisms at work) were consigned to oblivion. In contemporary urbanism, therefore, the emphasis is on increasing urban density and making city form subordinate to the efficiency of transport systems or to returns on land investments, as if the profound urban remodelling that this would require were a goal that could be achieved by democratic public policies. Politics has become imaginary but imagination in science is a victim of the "quantitative rationalisation" ([28]: 122) that favours prediction over explanation. There is still hope because, like the capacity for action of politicians, the predictive capacities of the social sciences vis-à-vis the urban are still very limited. Yet there is still a need for explanation. With hindsight, I continue to think that there were multiple misunderstandings around structural geography that could be corrected. Numerous avenues could have been explored, whereas there were others of which perhaps too much was expected. Let us therefore return to the trigger point. The quotation that follows sums up the scientific advances that emerged from this encounter and the hopes they nourished. It is taken from the preface that J. Petitot wrote for the book by G. Desmarais published in 1995, based on his thesis, "The Morphogenesis of Paris":

> The problems of morphogenesis can only be handled seriously if one has appropriate theories. In his work, Gaëtan Desmarais adopts and unifies three fundamental theories.
>
> (i) Gilles Ritchot's theory of urban form: this enables him to explain how the fabric of architectural forms is geographically connected by abstract morphological structures;
> (ii) Algirdas Julien Greimas's semiotic theory: this enables him to understand how meanings that possess a basic anthropological content [...] can play a constructive role in processes of territorial appropriation, in other words how spatial organisation can spatialise symbolic values;
> (iii) René Thom's morphodynamic theory: this enables him to link his research to a general mathematical theory of the genesis of morphological structures.

[3] https://www.erudit.org/fr/revues/cgq/1992-v36-n98-cgq2670.
https://www.erudit.org/fr/revues/cgq/1998-v42-n117-cgq2689/.

> Furnished with the three theories—which broadly fit into the scientific framework of structuralism [...]—Gaëtan Desmarais develops a thesis of rare interest. Like any compelling thesis, it is of course partly problematic and must therefore be carefully assessed and debated. But its importance to the understanding of the geographical essence of human settlement is such that it merits attention and consideration ([17]: 11–12[4])

In the 1970 and 1980s, structural approaches were used to analyse urban, architectural or geographical forms. By analogy with linguistics, the focus was above all on two types of "vocabulary". On the one hand, the vocabulary of the architectural forms or distribution of buildings, as signifiers of social relations, which prompted authors like Choay [5] to argue for history to play a part and hence for a departure from Saussurian structuralism. On the other hand, the vocabulary of geographical forms as spatial models handled by the language of maps, which Brunet [4] systematised in a grammar that he called *chorematic*. However, these approaches focused on just one kind of forms, whereas there are connections to be found between the objects contained in these hermetic terminological silos. This is what G. Ritchot managed to do.

The big problem with which this morphogenesis confronts the human sciences is that the form of the city, as represented by cartography—as an object in a so-called "geographical" space—is produced by social individuals for whom the city, as an element of human settlement as a whole, is the space in which they live. Within this space, they try—either collectively or individually—to increase their well-being, to control their environment and access to the resources they need, while accommodating to other occupants who share the same space. These are the internal dynamics of human settlement that are spatialised, that is to say externalised in forms that are not directly observable by social actors, but which guide them as agents and channel them as dynamic forces. A key issue, therefore, is to understand how the space in which these internal dynamics operate is structured.

6.1 G. Ritchot's Analogical Approach: From Geomorphology to Human Geography

G. Ritchot devoted two books to the theory of structural geomorphology, in 1975 and in 1984, the second cowritten with P. Laplante [13]. For its part, the theory of urban form was propounded in four publications: in 1977 in a report for the Québec Ministry of Cultural affairs [21],[5] in 1985 as a chapter in a multi-author book, in 1991 in a publication by his research unit that also includes a chapter on geomorphology [23, p. 183–210],[6] and in 2000 in a book cowritten with G. Desmarais. He explained on

[4] It is also in these terms that J. Petitot's contribution to the thesis defence in 1993 began.

[5] The report can be found online at http://liremaville.com, a website that aims to teach the interpretation of urban form, created by two urban historians and geographers from Québec, G. Gaudreau and N. Guilbault. Our thanks to them: http://liremaville.com/documents/rapport.pdf.

[6] Centre de recherche en aménagement et développement (CRAD) at Université Laval, Québec.

multiple occasions that the two theories are profoundly interconnected. In researching Montréal's architectural heritage in 1976 and 1977, G. Ritchot said that he pursued "an approach by analogy: the question of active form was first raised in structural geomorphology before being applied to the field of urban morphology" ([22], p.23). In order to understand the sources of this analogy, we will risk presenting these two complex theories in broad outline. In both cases, the theory seeks to answer a question raised by fieldwork that again I am unfortunately unable to present with the precision I would like.

1.a **In geomorphology…**

The fissures running through the rocky substrate can be grouped and classified into several morphological types. They partly explain the production of geological relief, but they cannot be explained themselves without accounting for the effect of a change of form on the substrate. In other words, explaining a form by forces always requires reference to other forms that have channelled those forces. It is the sequence that demands theoretical explanation. To this end, G. Ritchot imagines an undulating form that envelops the inelastic rock of the Earth's crust within a field of internal pressures. In isostatic equilibrium on the fluid magma, the tectonic plates can tip, lift or sink. Under the effects of lateral pressures, the rock formations are compressed in concave zones, ripped apart in convex zones or stacked in terraces that follow the inclined profiles. These compressions, rips or steps in the inelastic substrate can be identified empirically by faults, fractures, joints or grinding lines that vary in depth and extent. These phenomena are linked with "geomorphological fields" ([20]: 7), i.e. specific domains of erosion which, depending on the nature of the rocks (friable, plastic, brittle…) and the nature of their geological structure (dips in sedimentary layers, crystalline rock intrusions…) give rise to reliefs (plateaus, valleys, hills …) and characteristic facies (sand, scree…). According to G. Ritchot, therefore, it is not erosive weather agents alone which, by means of specific gravitational potential, cause the emergence of form by exhuming the hardest rocks and breaking down the softer rock. There is an abstract "primordial surface" which "traps" the erosive forces via a "structural" surface (in the geological sense of the spatial distribution of the rocks that constitute the substrate), resulting in a stabilised "erosion surface".

However, identifying this primordial surface is no simple matter and brings several hypotheses into play. G. Ritchot, for example, explains that several primordial surfaces can be cited to explain the formation of umbilics and rock bars in a glacial valley ([20]: 277). The correct one can be identified by analysis of the fissure patterns. However, the demonstrative value of this explanation depends on the structural stability of this primordial surface, which governs the long-term channelling of the erosive forces.

The undulations of the primordial surface are thus part of a hermeneutics in which the aim is to interpret the formation of land relief. For G. Ritchot, the primordial surface contributes to the construction of a *text* on existing forms, a text that lays claim initially to an operational, and subsequently to a scientific validity. This surface takes hold of the different domains and then subjects them to processes of erosion that destroy it.

> One form produces other forms at the cost of its own extinction: it is resolved into an abstraction and produces only a logical content through which we can understand the style of assembly of physical forms. But in practice the form-cause is lacking: if this is the status of urban form in general, it is a lost cause ([23], p.26)
>
> [...] structural geomorphology produces an algebra that suggests that forms can take the place of the active subject in the sentence. Forms can therefore be conceived as potentially causal, but consequently—instantly and perpetually—abstract (*ibid.*)

As far back as 1975, in order to explain the origin of this field of abstract folding that is the source of morphogenetic potential, G. Ritchot imagined a system of continental scale foothills in dynamic equilibrium between the tectonic plates (*ibid.* 151–164). This model, it has to be said, violates Wegener's tectonic plate theory. Although it accepts the initial interlocking of the continental plates, it posits not that these plates drift apart but that they are pushed apart by an increase in the volume of magma and therefore in the diameter of the earth. This controversy was scarcely calculated to encourage the diffusion of structural theory. Nonetheless, although the continental model was to be ultimately refuted, that does not invalidate the general approach. Local conjectures can remain plausible and above all pragmatically valid to serve as the basis for other hypotheses. As far as I can judge, this situation is common currency in the sciences that provide human or physical geography with reductionist models.

1.b **In human geography...**

En 1976, G. Ritchot headed a team of researchers commissioned by Québec's Ministry of Cultural Affairs to study Montréal's built heritage. They were interested by the way that certain urban forms (such as buildings, blocks, squares...) retained their value despite being functionally obsolete and abandoned by their former occupants. Such architectural forms retain exchange value and are successfully reoccupied by new actors, but their cultural value depends on their position in the spatial organisation of the city. Here again, the value of a form depends on another form. The economic explanation of rent (exchange value, Marxist labour value, Ricardo's differential value) is inadequate and the culturalist explanation (attachment to markers that depend on cultural schemas) disputable, because these attachments can clearly be identified with a dominant social class. Gilles Lavigne, another member of the team, suggested that rent should be understood as an "guaranteed positional income" ([14]: 123) that promises a better return on investment and increases the capacity to take on debt for a building project. In this view, it provides an incentive to erect buildings in some places that are more luxurious, more monumental, better made than their function requires, and conversely, in other places, structures that are cheaper and depreciate more rapidly. Guaranteed positional value also differentiates between practices and social groups. It encourages people to take up residence in prestigious locations in the city, i.e. places that are desired by and competed over by several actors. It is a feedback loop: competition increases the value of the locations and selects the urban functions that will convert this rise in value into income, without which the result would be no more than a speculative bubble. All this can trigger a

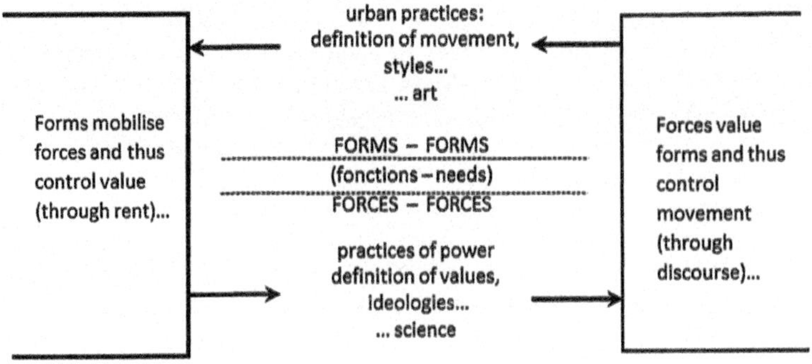

Fig. 6.1 Forms and forces schema 7 [22] (Adapted from Ritchot [22], p. 47)

cycle that lasts as long as the parties that own the places have the power to control the movements of the forces that give value to the forms (Fig. 6.1).

According to this interpretation, therefore, even in a context as capitalist as North America, the location of buildings and their functions is equally influenced by a longer term phenomenon than the activities of investors, a phenomenon that is largely symbolic. G. Desmarais read it as the outcome of a *generative path*, as defined by Greimas, which spatialises abstract values. These values evolve into political and economic meanings through two conversion operators that specify localities: on the one hand, the "rule of property" [6] that prevents the direct and predatory use of land and introduces political mediation, and on the other hand occupancy rights that resolve conflicts of appropriation and authorise an economic purpose. This path connects "three layers of spatiality" ([8]: 63):

i. "The deep layer consists in the anthropological dynamic of investment in values
ii. The middle layer is the product of a dynamic of political appropriation
iii. The surface layer relies on a dynamic of occupancy that is economic in nature

> These three layers are the spaces that anchor the internal dynamics, which act as "conversion operators" that effect the transitions between layers" (*ibid.*)

The theory of urban form thus situates the morphogenesis of human settlement within the perspective of a long period during which human communities laid claim to territories and divided them into subunits: administrative divisions and subdivisions or plots subject to the exercise of constantly changing property laws. I will return to this below. By contrast with geomorphological theory, which works with increasingly vast and all-encompassing forms, the theory of urban form moves back in time towards the production of very ancient spaces endowed with foundational values. Here, we approach the notion of the sacred. G. Ritchot identifies the "vacuum" of *Les Innocents* in Paris, the cemetery over which the *Les Halles* food market used to stand between the 12 and 18th centuries, as the theatre of salvation for bourgeois souls through the intercession of the Holy Innocents. The Catholic Church had traditionally inhibited the development of markets in cities through its condemnation of usury, so this place marks the advent of a new era that would last several centuries.

G. Desmarais reinforces the notion of the *sacred vacuum*, drawing on R. Girard's anthropological theory to explain the foundational role of the Plain of Lendit—a sacred and forbidden place where the town of Saint-Denis stands today—in the development of Lutetia, the city that became Paris. Such forbidden places symbolise the protection from neighbouring settlements afforded by the tutelary deity, but also the fear of his rage which, as with all pagan divinities, can be turned against his worshippers. For Girard, this is the ambivalent nature of the emissary victim—the scapegoat: a protective or unifying figure after the sacrifice but—or because—the repository of all evils before departing this world. This, in his view, is the hidden meaning of the victimary violence invested in the "vacuum" of the Plain of Lendit: a place where human settlement was forbidden, but which conferred value on the places that controlled access to it and hence on the buildings erected there, with the result that they subsequently profited from the presence of the celebrated Fair of Lendit. In this account, the sacred vacuum constitutes the model for the origin of human settlements.

Like the continental drift theory, this notion attracted criticism, notably for its pretensions to explanatory universality, but this is not the essential component of the structural explanation.[7] In his talk at the *Actualité de René Thom* symposium, Ch. Bayle showed that a less general and less ambitious model of the symbolic occupancy of a space previously emptied of its physical occupancy was entirely sufficient for an architect and urbanist like him—planner of the Paris Rive Gauche district at SEMAPA—to reconstruct and understand the operational effectiveness of his projects. Ultimately, the most important point for me is that a vacuum is a spatial configuration which embodies a relationship with an *asymmetrical Other* who, by the sole power of his name, prevents any actual occupancy although in practice making no use of the space. The theory of urban form is thus a way to introduce an assumed and spatially embedded principle of asymmetrical otherness into the morphogenetic process of human settlement. In consequence, it avoids the attribution of a theoretically symmetrical otherness between individuals, noting that that this symmetry would have been betrayed by social conventions or power relations produced independently of space. Human settlement thus organises and structures the social otherness that subjects encounter in their spatial trajectories. I will return later to this profoundly structuralist point.

[7] G. Ritchot proposes a third, more temporary, sort of sacred vacuum, which regulates the warring relations between Amerindian tribes [24]. This illustrates the great capacity of vacuum theory to adjust to the facts. To qualify the role of the vacuum, moreover, it should not be forgotten that the values invested are then concealed by actual buildings and spread beyond this space.

6.2 The Strengths of the Analogy

The analogy can be developed in three steps. It draws a comparison first between two dynamic processes: the erosion of landforms and the erection of human settlements (i.e. the production of architectural forms and anthropised landscapes); second, between two geometries: the geometry of faults and fissures in the bedrock and of property and territorial boundaries; and third, between two abstract and theoretical morphological structures. Let us start by stating that the unrestricted action of the forces at work in the processes of erosion or erection would lead, according to the theory of forces, respectively to the disappearance of the forms of relief or of urban settlements. As a consequence, since such theory is not entirely false, i.e. is locally true, it can be supposed that these mechanisms are spatially contained within domains where their external conditions of operation are durably differentiated. We then need to explore how these domains can be generated by a continuous and stable primordial form that envelops the discontinuous substrate.

There are nevertheless differences. First, erosion produces forms by removing material, whereas erection produces them by adding it. Then, erosion uses a physical potential, whereas erection has to draw physical energy from nature in order to give real content to a potential of symbolic value. In consequence, the morphogenesis of human settlement cannot ultimately be separated from ecological considerations. Next, the big difference is that the internal mechanisms of destruction of potential are, in the one case, physio-chemical, the same everywhere on earth and unchanged since the dawn of time, and in the other case politico-economic, developed within local cultural systems that have historically been in constant evolution, and that vary in their extension and influence within terrestrial space. The discontinuities are therefore also radically different in kind, hence the use of the term "fissuring" to refer to those in the geological bedrock and of "partitions" for the symbolic discontinuities in human settlement.

2.a **Spatialising a general homogenising mechanism**

If local forces are left to themselves so that their effect can spread gradually across the whole earth, the states they ultimately produce are amorphous. This point was theorised in relation to geological relief by W.M. Davis in the late 19th century with the notion of the *peneplain*, a surface that remains completely stable until the next orogeny. The process that plays the role of erosion for human settlements is *erection*. In fact, this is a deceptive term, because it suggests verticality, whereas the production of horizontal forms such as gardens, fields, storage areas or transport infrastructures also contribute to the process. Erected forms accommodate and locate human activities. They fulfil functions with a degree of efficiency. Some geographical positions are more favourable than others because they provide better access to factors that are useful to these activities: this local advantage is a differential guaranteed income (distinct from the type identified by Lavigne). In a territory without guaranteed income, the decision to allocate a space to a given economic function would be indifferent to position. It would not produce a flat surface, but a random human settlement, reflecting the Brownian movements of the occupants. In this regard, Rebour [18] drew attention to Heckscher, Ohlin and Samuelson's theorem, the standard model of international trade. He showed that, if generally applied, this model would theoretically culminate in the conditions of production becoming spatially uniform through the disappearance of differential incomes. This is because low wages would be offset by lower productivity; migrations of individuals to high wage areas would result in the means of production moving in the opposite direction, ultimately preventing the accumulation of people and capital in the same place, and hence the development of cities. It is the partitioning of the world that makes it possible to lock off income flows and to create accumulations of value. Capital must constantly be added to urban positions, by taking on debt, in order to prevent them dispersing and diluting their power to attract dynamic populations. True, the analogy is limited by the fact that economic mechanisms have none of the universality of physics. However, with the generalisation and globalisation of markets, it seems likely that this mechanism of homogenisation has been underway in some places for several centuries, and in others for several decades.

To sum up: the existence of forms in relief or human settlement arises because these processes, which are "normal" under the laws of physics or economics, can be countered by means of a partitioning of space that differentiates a substrate in domains positioned in higher-scale morphological structures, or forms.[8] However, this positioning is not mechanically determined. Locally, the pattern of the plot borders and territorial boundaries, or of the faults and fissures, is largely a product of chance. It carries the traces of old events that have marked it to different degrees. However, the probability of fissures occurring in a rocky substrate in a particular region or a territory being divided between different owners can depend on higher-scale—continental or geopolitical—force fields, which may reactivate ancient discontinuities. These are far from parasitical force fields, they are fields that produce forms. So we also need to understand what it is that regulates this production of discontinuities and prevents the substrates atomising, which would lead to homogenisation by another path.

2.b Positing the genuine discontinuity of geographical space as the basis of all possible knowledge about morphogenesis

The reality of the discontinuity of the rocky substrate in geomorphology is beyond dispute. Geological faults are observable and irrefutable realities. The same is true in human geography. The partitioning of the world is unquestionably a reality and a fundamental property of geographical space, as Gottmann [11] wrote. However, this partitioning is symbolic in nature, whereas building is a physical transformation of the world. Across the materialist tradition, the importance of partitioning tends to be underplayed and its permanence denied. But it is not just a symbol like any other. Its is founded on the principle of ownership (first collective, then individual ownership), a principle that is also about recognition of the *name*, as G. Mercier

[8] It should also be noted that the systems subject to these so-called "normal" processes are gradient systems, and therefore systems where Catastrophe Theory can be used to explain the direction of rises in potential.

notes ([16]: 91). The name represents the *other* in the sense that it occupies a space and by its presence prevents any other occupancy. It also gives the *other* a symbolic presence that is equally exclusionary. Partitioning thus contributes to the existence and even the emergence of the *other* as a being capable of imposing his rationale. This theme of the "other" (in which Ritchot drew on his reading of Freud and Lacan) was fundamental in the theory of urban form as formulated in 1977.

The application of property rights is therefore implacable, despite its variations, whether through urban planning rules at the scale of small units, special arrangements for areas of strategic interest, or fundamental legal or even constitutional principles at the scale of the biggest spatial units such as countries. While the analysis of partitioning long focused on frontiers at the geopolitical scale, there is nothing to prevent methodological development which would extend it to all the boundaries produced by the law over time. Today, for example, we have digital land registers, including reconstructions of historical land registers. Outside societies with land registers, and there are many of them, the pattern of these discontinuities can be uncertain but, while qualitative differences between positions are clearly acknowledged, it does not much matter where precisely a frontier runs. On the other hand, the evolution of a vague partition should differ from the boundaries in in a land register.

Any occupation of space excludes all other occupancies and therefore creates discontinuities for a time. This needs to be stressed, since it is an essential geographical property.[9] The collective organisation of this occupancy thus entails control of the production of these discontinuities and, more importantly, of their erasure. The fissuring mechanism in geology and the partitioning of the world in geography are two spatial objects. They are therefore geometrical varieties, i.e. spaces, which stand between the rocky substrate and the agents of erosion, or between nature and its exploitation by human settlements. These discontinuous, dynamic and potentially unstable spaces are therefore the empirical basis of G. Ritchot's structuralism. Qualitatively different domains are distinguished and connected together by relations that bring to mind the relations that organise the words in a language.

This is a profoundly original approach in geography, indeed I have no hesitation in describing it as revolutionary. Yet this analogy, which explains the role of symbolic partitioning and its geometrical organisation in the dynamics of the substrate seems has gone unnoticed. In any case, it has been little emphasised among the disciples of G. Ritchot despite the pioneering work of Guenet [10] on the suburbs of the Upper Town of Quebec, or perhaps because Guenet showed no statistical correlation between the plots morphometry and the types of settlement trajectories that generated them—but the relationship cannot be that simple. Today, it is tempting to see this originality as one of the reasons for the incomprehension with which structural geography has often been met. It requires a change to the fundamental referent of knowledge: no longer the most stable—supposedly the most "objective"—forms described within a continuous space, but a discontinuous and unstable space covered

[9] An assertion that can be found among several geographers who have contributed to the epistemology of the discipline, for example H. Reymond or G. Nicolas.

by shapes that may "change, as we know, more quickly than the mortal heart."[10] In the first case, science is challenged to explain what it is that sets this cumbersome theatre in motion and how inflections or bifurcations cause it to depart from the simple reproduction of its structures. In the second case, science is faced with discontinuous spaces tragically threatened with disintegration and the first theoretical act in attempting to understand their historical dynamic is to posit the structure that keeps them stable.

6.3 Complexity and Specificities of Abstract Forms in Human Geography

Identifying the primordial surfaces is the key to explaining morphogenesis, and partitioning is the way it acts on physical forms. But these surfaces do not exist empirically. Envisioning them requires a very good knowledge of the mechanisms which, at different scales, contribute to the emergence and therefore to the differentiations of the internal dynamics. As I sorely lack the necessary knowledge in physical geography, from now on I will confine myself to human geography. By its reversal of perspective—sadly misunderstood as it is—the theory of urban form has, it seems to me, brought new capacity to explaining the complexity of human settlement.

3.a **Partitioning of space and positioning of otherness**

The process of morphogenesis in human settlement operates on at least three levels.[11] We will employ the word "layer" to refer to the space where the internal dynamics take place and the word "level" to refer to the set of forms that constitute the external product of these dynamics and gain meaning from them. A hasty reading of the process, which was my case for a long time, may suggest that meaning can only be grasped at the level of the form of the surfaces, i.e. of the built landscape and its architectural codes. However, meaning is in fact produced and grasped at every level, because the separation between the dynamic layers is theoretical; it is devised to help understand the causal complexity of human settlement. The habitual space of geographers encompasses all of them together. The theory therefore envisages that a dynamic in a certain layer uses the lower-level signals to produce those of the upper level, with the possibility of feedback. In this way, land transactions produce a localised meaning that emerges from a layer of economic dynamics. The decisions to allocate a plot of land to one of the big social activities, such as manufacturing, housing and family life, farming, administration, the army, religion... are still the meaningful forms of an underlying layer of political dynamics, where the involvement of economic actors imposes the requirement of profit.

[10] The first sentence of the essay by J. Gracq, "The Shape of a City" which refers to a line by Beaudelaire in the poem "The Swan", evoking the changes to Paris brought about by Haussmann.

[11] There are also several levels in geomorphology (see "structural" and "erosion" surfaces above).

Each level selects a system of signs, a way of imparting meaning to positional differences. Individuals respond to them according to their goals, and according to the preparation they have received through their upbringing and education. The surface level can appear to be the most monotonous and routine level or conversely the one that most teems with signs for imaginations attracted to freedom. In *The Shape of a City*, Gracq [12] describes Sunday strolls in the outskirts of Nantes which led him, a boarder at a secondary school, towards patches of wasteland: "dream spaces as well as areas to roam free". He describes himself as struck by "the particular capacity of a city to supply a poetry-loving imagination indefinitely and flexibly with the markers of models and paths, to *give onto* the most original visions almost naturally, without the need to be forced in any way" (*ibid*.: 64). The surface level is a landscape so saturated with signs that the adult individual often ceases to notice it. However, the child opens up to this world of meanings and, through his egocentricity and imagination, easily puts himself in the place of the other if he feels that the other has departed.

Meaningful objects in the built landscape generally reveal the presence of the other and the other's occupancy, guaranteed by territorial partitioning. The houses on their plots, the place of worship identified with its community, the social housing block, the streets where traffic is governed by the public authorities… illustrate the diversity of the forms of otherness. Through its hierarchy and its differentiation, partitioning brings out a geometry of otherness. It conditions the realisation of a subject's settlement trajectories, since it determines the types of social actors that he or she will encounter, as well as the asymetrical modalities of encounter. For example, the trajectory will be radically different if the space is divided into a multitude of small plots with privately owned dwellings or only a few estates rented out as farm plots. One could argue that at the moment of their emergence, there is a dual correspondence between forms of spatial partitioning and forms of social otherness. As law evolves, these forms can then take on separate—and potentially conflicting legal existences.

3.b **Paths of meaning and time inversions**

However, these signs that create the link with otherness do not all belong to the same causal registers. "Viewed from the perspective of the connection between past, present and future, time poses a dual problem when it comes to spatial planning: the problem of the influence of legacies on the content and structures of space, the problem of the causal relations between present and future", writes Fischer [9], whom I was lucky to have as my thesis supervisor. The surface level shows the established occupancies. The present of the signs is determined by their past, by the durability of the buildings, by the profitability and use of the services, by the validity of the contract, by the inertia of traditions. However, otherness changes suddenly as a result of property deals or the signature of rental leases, though the partitions remain the same. Each of these events generates a price that activates semaphores on a plot. The economic signals respond in the short timeframe of market adjustment, because individuals who want to find a home or invest live in the immediate present. Finally, at political level, the decisions to dedicate particular places to collective activities through which the community hopes to perpetuate itself and improve its conditions of existence, take on an air of solemnity and elitism. They are associated with an intervention on territorial division or, at the very least, with the arrival of a new occupant at the end of a specific trajectory, such as the acquisition of farm plots by a railway company. Here, it is the future that determines the present.

The vertical connection between layers and levels thus depends on partitioning, which is itself horizontally structured and scalable. The relationship between layers is not one of causal determination—because the ultimate causality of the political level cannot influence the balance between supply and demand nor actual causality at the surface level—but of *overdetermination*. Thus the styles and sizes of the buildings are the product of interactions that are stimulated by several factors that are significant at the economic level. Likewise, the meaning of these transactions is elaborated in the internal dynamics of the appropriation plans, which may be complementary or conflictual at the political level. Beneath this stack is a further, more collective and cultural—and in this sense *identitarian*—layer "of investment of profound anthropological values in places" ([8]: 62), which in turn overdetermines the political level. There are at least two reasons for this. One is speculative and logical in kind, since the discourse used to explain the form of settlement needs to posit poles of attraction and repulsion that steer the original settlement trajectories of individuals. The other reason is more fundamental—ontological in kind?—since the reality of symbolic partitioning, insofar as it inhibits the immediate satisfaction of any human being's needs or desires within a space transformed into a territory, governs the possibility of sociability for those same human beings. Any object of partition would thus bear the trace of a "dismembered transcendence" [27], here that of the asymmetrical other. The role of property thus takes on the status of an anthropological prohibition.

6.4 Work in Progress in Structural Human Geography

A final look at the analogy with geomorphology reveals another difference between the two theoretical approaches. Whereas the undulations of the primordial surface might partially determine the morphology of the fissural mechanism by means of a few physical laws, the abstract form of value investments and the morphogenetic path tells us little about the morphology of symbolic partitioning. The dynamics of land and administrative divisions seem to have been left out of the theory,[12] as have the forms—in this case not socio-spatial but socio-temporal—instantiated by contracts of territorial appropriation or occupancy. Yet conflicts between agents are typically resolved by means of contracts: one of the two parties maintains a position and its

[12] On the other hand, those dynamics have not been forgotten in studies, for example the study on the morphogenesis of Québec [24, 25], which assigns all the necessary importance to provincial subdivisions, from the French defeat to the establishment of the Canadian Confederation in 1867, or to the maintenance of seigneurial tenure in the Lower Saint-Lawrence region. Nor is it ignored in Th. Rebour's théorie du Rachat [19]. However, the determinism with which the latter theory operates in the economic layer makes it difficult to connect it with the theory of urban form, despite the obvious kinship.

values but makes the other party (here rendered subordinate) accept or even desire a position in return for a restriction on its mobility: "the control of movement by speech", as G. Ritchot wrote in 1985 (Fig. 6.1).

A great merit of a semiotic analysis of the phase of conflict for territorial appropriation has, in my opinion, been to highlight a family of dissymmetrical contracts, the purpose of which is to grant the weaker contracting party access to a portion of space from which, at the cost of restrictions on its mobility, it can hope to obtain protection, safety or subsistence, in other words a quantum of inherently desirable time. Even the original mythical settlement, situated a good distance from the *vacuum*, can be interpreted as a contract established by a human group with a transcendent and protective power, in accordance with R. Girard's theory of the emissary victim. As for the hoped-for future ecological and sustainable human settlement—unless it is also mythic?—does it not demand the establishment of a broad "natural contract", to use the expression coined by M. Serres, if it is to advance beyond the status of utopia?

Very promising avenues are therefore open for future research, such as studying the link between the formation of social denominations and the evolution of spatial partitioning via the production of laws, employing the cusp catastrophe model as in the case of Marcos ([15]: 6), or analysing the genealogy of the relations within a fundamental division in society, that between the possessors of knowledge, of power, and commercial intermediaries ([3]: 242). However, actantial simplification seems to be limited to the art of interpretation according to a static framework, in keeping with a much criticised structuralist tradition. Yet it would be of the greatest interest to explore the political, economic and technical context within which a social actor adopts a given actantial role, and the contractual form this might take.

Despite the contacts made and intentions expressed, the applications of R. Thom's ideas remain concentrated within the field of semiotics thanks to the theory of *prégnance* and *saillance* and seem rather rare or shallow in the field of mathematical modelling. They ignored genuinely quantifiable and gradual dynamics, for example social differentiations, the evolution of spatial partitioning or the development of the market and the formation of prices, which should be suited to mathematisation. This point has been explored recently in a study that I will now approach from that angle.

O. Bonin and I tried to model a highly simplified morphogenetic process by focusing on the economic level internal dynamics in land transactions [2]. The model contains significant simplifications: it envisages an endogenously growing population; on reaching adulthood, young households randomly but configurably receive a certain income that they use to acquire a dwelling in a competitive real estate market; there is no vacuum: the dwellings are built on an empty plot; plot divisions are fixed, as is the location of jobs, and the urban planning rules are uniform; the legal mode of allocation or occupancy may equally be rent or sale, the purchase price being converted into a fictional rent. Households will therefore occupy positions according to a balance between income, housing costs, transport costs for travel to work, and their aspirations for a large dwelling and access to as many local amenities as possible.

The allocation of a dwelling is calculated using the standard model for the formation of real estate prices in urban economics, developed by Alonso, Mills and Muth in the late 1960s. According to this model, urban form tends to be organised in concentric circles with no break in radial symmetry: jobs in the centre (their initial location), the wealthy occupy big houses on the outskirts because their cars give them quick access to the jobs in the centre; and the poor remain in an intermediate zone to reduce the costs of transport to the centre. However, the auction mechanism that sets land prices does not take into account the possibility for occupants to enhance value, a potential that has the effect of polarising certain trajectories of residential mobility. Our model accounts for this by endowing some of the households with "sensitivity" to this value, which will be converted into positional income. It reflects the fact that households want to take action to increase local amenities, invest in the quality of their neighbourhood and possibly receive income for their activities or exploit their assets, as in G. Lavigne's model. The *value* in a place therefore increases with the number of households that settle there and are sensitive to it. This *quantum* of added value varies according to the household's revenue. The model then brings into play a mechanism of attraction, or value self-catalysis, so that the sensitive households gather around clumps of value and make them grow. This mechanism makes them accept a less spacious home in places where the positional value is high. They see themselves as ultimate beneficiaries even though, in the eyes of the other households, the sacrifices they make to obtain these smaller dwellings do not seem worthwhile.

It would therefore seem that introducing this factor of household sensitivity to a symbolism implicitly present in urban forms, revealing advantages that do not yet (and perhaps will never) exist, alters the spatial distribution of buildings and income, and breaks the radioconcentric model proposed by Alonso, Mills and Muth.[13] The elevation surface of positional value is shaped like a threshold, like the one reconstructed by the analysis of morphogenesis in Montréal [24, 25] or Paris ([23]; Fig. 6.2).

It goes without saying that the model does not reconstruct the formation of the positional value thresholds in Paris in the 13th century or Montréal in the 19th. Nevertheless, it shows that the threshold is a generic configuration in the same way as the cone of revolution, and that it develops when households have different ways of evaluating their expectations of the positions they wish to occupy.

All the glaring failings of this model are potential avenues for research leading to multiple intersections where Thom's approaches may be encountered, be it in mathematics, semiotics or philosophy.

[13] As J. Petitot pointed out in a seminar at the Ecole des Ponts in October 2012, we followed a Turing type approach: an enabling "morphogene" self-catalyses following a relatively gradual dynamic—positional value "unbalanced" by guaranteed income—whereas other inhibiting morphogenes have a faster dynamic (the availability of dwellings at an affordable price). These were roughly the conditions proposed by A. Turing in a famous 1951 article.

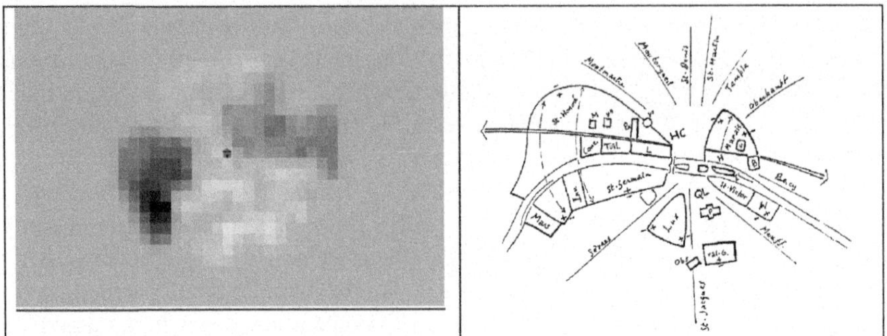

Results of simulation from a minimum city nucleus, with constant jobs. We see the formation of an abstract relief of positional value, © Bonin, Hubert 2023

Ritchot 1991, *Sketch 8: morphological configuration of Paris around the end of the Ancien Régime.* (Reproduced with the permission of the CRAD)

Fig. 6.2 Morphologies of urban thresholds, obtained by simulation or urban form theory

Acknowledgements I would like to thank Isabel Marcos and Clément Morier who, through their seminar *Actualité René Thom* at Université Paris-Diderot, gave me the opportunity to revisit these structural theories and to exchange for the last time with Gilles Ritchot in October 2019, on the construction of this article. I also thank Olivier Bonin for his helpful input in the role of proofreader and Luc Ampleman for his help to complete the list of doctoral dissertations in human structural geography.

References

1. Ampleman, L.: Transport Geopolitics. Decoding and Understanding Transport as a source of Conflicts. Palgrave Macmillan, Singapore (2021). https://doi.org/10.1007/978-981-16-4967-7
2. Bonin, O., Hubert, J.-P.: Modélisation morphogénétique de moyen terme des villes: une schématisation du modèle théorique de Ritchot et Desmarais dans le cadre du modèle standard de l'économie urbaine. RERU, Armand Colin **2014**(3), 471–497. https://doi.org/10.3917/reru.143.0471. (2014)
3. Brandt, P.A.: What's new?—50.000 years of Modernism. A semiotic rhapsody. In: Spaces, Domains, and Meaning. Essays in Cognitive Semiotics, European Semiotics Series, vol. 4, pp. 234–256. Peter Lang (2004)
4. La, B.R.: Composition des modèles dans l'analyse spatiale. L'Espace géographique, tome **9**(4), 253–265 (1980). https://doi.org/10.3406/spgeo.1980.3572
5. Choay, F.: Sémiologie et urbanisme. In: Choay, F., Banham, R. (dir.) Le sens de la ville. Le Seuil, Paris (1972)
6. Desmarais, G.: The Sacred Place: a morphodynamic hypothesis for the foundation of human settlement. Nordisk Arkitekturforskning **8**(4), 39–50. http://arkitekturforskning.net/na/article/view/674 (1995). (Semiotik)
7. Desmarais, G., Ritchot, G: La modélisaton dynamique en géographie humaine. Numéro special. Cahiers de géographie du Québec **42**(117) (1998)
8. Desmarais, G., Ritchot, G.: La géographie structurale. L'Harmattan, Paris (2000)
9. Fischer, A.: Temps et aménagement du territoire. L'Espace géographique, tome **8**(2), 81–84 (1979). https://doi.org/10.3406/spgeo.1979.1876

10. Guenet, M.: Géographie structurale urbaine: expertises de cartographie automatique à Québec, thèse Ph.D., Géographie, Québec, Université Laval (1990)
11. Gottmann, J.: La politique des Etats et leur géographie, Paris, CTHS, 2007, préface et bibliographie complète des travaux de Jean Gottmann par Luca Muscara (1952)
12. Gracq, J.: La forme d'une ville. José Corti, Paris (1985)
13. Laplante, P., Ritchot, G.: La forme de la terre. Editions du Préambule, Longueuil (1984)
14. Lavigne, G.: La rente urbaine. In: Ritchot, G., Feltz, C. (eds.) Formes urbaines pratiques sociales, pp.101–131. CIACO Editions du Préambule, Longueuil Louvain-la-Neuve (1985)
15. Marcos, I.: Urban morphogenesis. Semiotica **2012**(192), 1–14. https://doi.org/10.1515/sem-2012-0077 (2012)
16. Mercier, G.: Propriété privée, crédit, travail. In: Ritchot, G., Feltz, C. (eds.) Formes urbaines pratiques sociales, pp. 71–97. CIACO Editions du Préambule, Longueuil Louvain-la-Neuve (1985)
17. Petitot, J.: Préface. In: Desmarais, G. (ed.) La morphogenèse de Paris des origines à la Révolution. L'Harmattan/CELAT, Paris/Sainte-Foy (1995)
18. Rebour, T.: La ville comme phénomène de géographie économique. Cahiers de géographie du Québec **41**(113), 201–207 (1997). https://doi.org/10.7202/022641ar
19. Rebour, T.: La théorie du Rachat. Publications de la Sorbonne, Paris (2000)
20. Ritchot, G.: Essais de géomorphologie structurale. Presses de l'Université Laval, Québec (1975)
21. Ritchot, G., Charbonneau, F., Gascon, P., Lavigne, G.: Rapport d'étude sur le patrimoine immobilier. Centre de recherches et d'innovation urbaines, Montréal (1977)
22. Ritchot, G.: Prémisses d'une théorie de la forme urbaine. In: Ritchot, G., Feltz, C. (eds.) Formes urbaines pratiques sociales, pp. 23–65. CIACO Editions du Préambule, Longueuil Louvain-la-Neuve (1985)
23. Ritchot, G.: Etudes de géographie structurale Cahier du CRAD n°15, (collaboration de G. Desmarais). Université Laval, Québec (1991)
24. Ritchot, G.: Québec, forme d'établissement. Etude de géographie régionale structurale. L'Harmattan, Paris (1999)
25. Ritchot, G.: Montréal: entre secret et mensonge. In: Marcos, I. (dir.) Dynamique de la ville. Essai de sémiotique de l'espace, pp.19–48. L'Harmattan, Paris (2007)
26. Ritchot, G., Mercier, G.: La géographie humaine structurale. Numéro special. Cahiers de géographie du Québec **36**(98) (1992)
27. Thom, R.: Postface. La Transcendance démembrée. In: Pinchard, B. (ed.) La raison dédoublée, pp. 575–596. Aubier, Paris (1992)
28. Virole, B.: Catastrophes de l'inconscient. Les éditions Baghera, Pantin (2019)

Chapter 7
Psychoanalysis and Thom's Semiotics in Dialogue: Territorial Psychoanalysis

Isabel Marcos and Carlos Farate

7.1 Introduction

The general idea of this chapter that sets the foundations of the interdisciplinary scientific field of Territorial Psychoanalysis is primarily inspired by authors such as Gilles Deleuze, Félix Guattari,[1] René Thom, Per Aage Brandt, Sigmund Freud, Jacques-Marie Lacan, and Bent Rosenbaum. Benjamin ([2]: 16), in his book "The Arcade Project: Passages" succinctly summarizes the founding idea of the morphogenesis of borders: "The city is only homogenous in appearance. Its name takes on a different accent depending on where you are. Nowhere—except in dreams—can one experience the phenomenon of the bordering as originally as in cities. To know them is to know where the lines that serve as demarcations run, along viaducts, through buildings, in the heart of the park, and along the riverbank; it is knowing these borders as well as the enclaves of different domains. The border cuts across streets; it is a threshold; one enters a new domain by taking a step into the void as if one had crossed an unseen step."

Our chapter titled "Psychanalysis and Thom's Semiotics in Dialogue: Territorial Psychoanalysis" presents Territorial Psychoanalysis through two different, whilst convergent, empirical case studies: a socio-urban case study (the morphogenesis

[1] Deleuze and Guattari were a source of inspiration for this « Territorial Psychoanalysis » project.

This work has been financed by national funds through FCT—Fundação para a Ciência e Tecnologia I.P.—within the framework of the R&D Unit Centre for Philosophy of Science of the University of Lisbon (CFCUL), a strategic project with the FCT IP. Reference: UIDB 00678/2020 and UIDP/00678/2020.

I. Marcos (✉)
University of Lisbon, Lisbon, Portugal
e-mail: p900292@ulusofona.pt

C. Farate
Miguel Torga Institute, Coimbra, Portugal

© The Author(s), under exclusive license to Springer Nature Switzerland AG 2024

I. Marcos and C. Morier (eds.), *The Relevance of René Thom*, Lecture Notes in Morphogenesis, https://doi.org/10.1007/978-3-031-54982-3_7

of boundaries that structure a specific territory) and an individual case study (the morphogenesis of the borders that structure the psychic territory of a human being). After the epistemology inherent to this new interdisciplinary field, we will first present the urban case study. This will allow us to examine the way the complex of the *Salpêtrière* hospital is influenced by its original traumatic inscription, in other words, how this space can be interpreted as a symptom of an imaginary of rejection by the city of Paris. This imaginary of rejection also works as an unknown and frightening instance circumscribed by the structuring lines of the urban space, defined by inner-outer border zones. This case study also explores the psychological importance of a traumatic experience, which, in this case, was removed from consciousness through a process of mnemic omission, akin to forgetting, and thus remained "crystallized" as implicit collective memory.

The second case study refers to the experience of psychic trauma endured by a young woman, whose traumatic experience took place in the same physical space of the *Salpêtrière* hospital. Therefore, the *Salpêtrière* hospital, both as socio-cultural emplacement rejected by the city of Paris and as the place of psychic trauma rejected by the young woman who experienced it, forms the territorial link that bridges the two case studies, serving, that way, as the pivotal point for the theoretical-conceptual dialogue between urban space and the human mind.

Such exchange between urban space and the human mind takes a dialogical form, in other words, takes the form of an intersubjective relationship between the inhabited territory and the subject residing in it. This implies the need for interdisciplinary exchange between Thom's semiotics and psychoanalysis to grasp the morphodynamic nature of the construction of both a collective urban territory and an individual inner territorial space. In this regard, Thom semiotics, founded by René Thom, a Fields Medal in mathematics (1958), also known for its influence on the epistemological development of morphodynamic semiotics, presents, according to Brandt [9], two semiotics: the theory of catastrophes and the theory of saliences and pregnancies.

These semiotic systems translate to mathematics a psychological problem as they seek to elucidate the relationship between signs and meaning, a relationship that lies at the heart of language as a privileged tool for thinking and understanding both the surrounding space and oneself [26].

This premise leads to a central demonstration: morphologies are universal and create the possibility of developing a science of forms with its own syntax and semantics, where the units are local elementary morphogenetic fields that can be modeled by Thom's elementary catastrophes [35]. This is why we will begin by elaborating on the concept of the morphogenetic field at the beginning of this article.

In its own right, psychoanalysis, as the science of the unconscious [17], developed philosophically as a field of speech and language by Lacan, offers topographical systems that aim to seek hidden, concealed psychic meanings in the conscious or brought-to-consciousness discourse of the subject, through the signs of everyday language, dreams, daydreams, as well as through the body and behavior, interpreted as (false) statements at the service of the subject of enunciation's "sinthome" [23].

Furthermore, psychoanalysis places, from the outset, significant importance on studies and research applied to other disciplinary fields, especially the humanities and the social sciences, often in the form of interdisciplinary studies. The objective of this chapter is precisely to establish the heuristics of the critical conjunction of psychoanalysis and Thom's semiotics as the epistemology of a new interdisciplinary field: Territorial Psychoanalysis. The common thread of this initial theoretical moment is the notion of border, simultaneously as a threshold and a separator, in the form of an internal and external intersubjective "double limit" [8, 19]. And this in the double psychodynamic (individual) and physical-symbolic (urban) dimensions.

Since this new interdisciplinary field offers a method for the morphodynamic analysis of urban construction based on the dialogical and intersubjective relationship between the inhabited territory and the subject who inhabits it, we have chosen the concept of trauma, viewed from a diachronic perspective that emphasizes the concepts of "deferred action" (*nachträglich*) [16], "après-coup" [12], or "cumulative trauma" [21] as the empirical-conceptual domain of our research.

Finally, Territorial Psychoanalysis borrows from Thom's dynamic semiotics the key concepts that will be deployed by the theoretical tools, the epistemological foundations, that define and delineate this new and innovative scientific field: Morphogenesis—Morphogenetic gradient, Vacuum, Symptom—Morphogenetic gradient, Border, Pregnancy—pregnant field, Salience.

7.2 Morphogenesis—Morphogenetic Gradient

We begin by unveiling the concepts of Morphogenesis and Morphogenetic Gradient,[2] originally applied to the urban space, in other words, to a collective space. Subsequently, we will proceed with the analogical application of these concepts to the individual space, specifically the psychic space of a woman suffering from post-traumatic stress, which is the human subject of the individual case study. There is also a symbolic link between the collective unconscious and the individual unconscious since this place, the *Salpêtrière*, is both the material and the imaginary-phantasmatic traumatic scenario for either the collective (urban) identity disorder or the individual (personal) identity disorder that simultaneously unites and separates the two protagonists.

7.2.1 The Salpêtrière *District: About Urban Space*

Currently, the *Salpêtrière* district consists of a hospital dating back to the 17th century, a 19th-century train station, both classified as historical monuments and a set of

[2] We thank the town planner responsible for this urban planning operation for giving us access to detailed information on this neighbourhood project and for the initial discussions for this analysis [30].

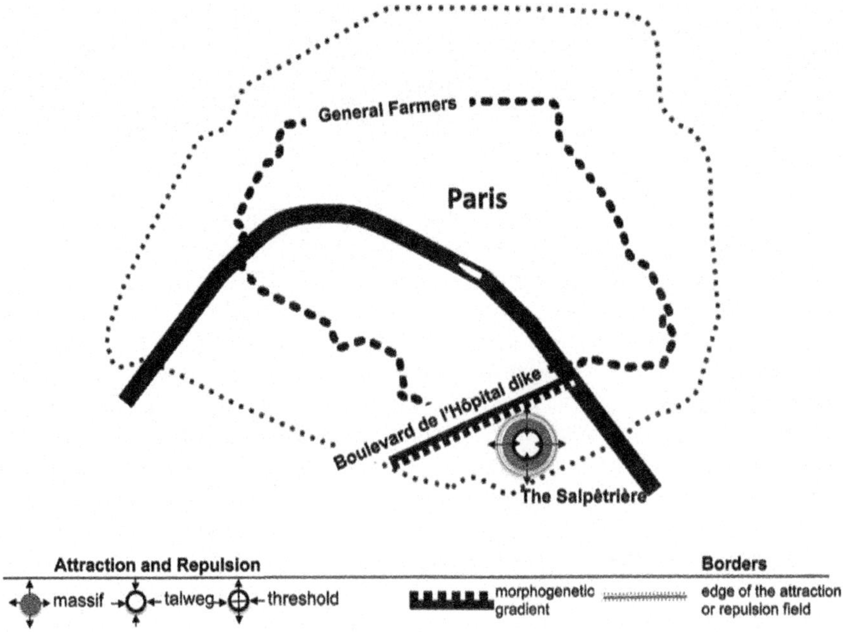

Fig. 7.1 The primitive dike of Boulevard de l'Hôpital, 17th Century. This dike represents a morphogenetic gradient, a dynamic of territorial appropriation that allows the study of mobility trajectories being attracted or repelled. This appropriation enables the actualization of urban values

"barracks" and interstitial spaces that challenge any urbanistic conception, resulting in a rather confusing urban interpretation of this district.

Following the hypothesis that we will develop in this section, the district is historically marked by a repulsive field, corresponding to a stable evolution.[3] Figures 7.1, 7.5, 7.6, 7.7, and 7.8 help us understand the repulsive field that is the *Salpêtrière* in its relationship with the city of Paris and are inspired by the representation of a dynamic system with two variables [18]. We turn to Gilmore because his model allows us to represent a field of attraction or repulsion with several separatrices, ridgelines, talweg lines, and passages.

[3] The Thom-Mather theorem, and subsequently, the theory of elementary catastrophes that applies this theorem, asserts that under certain conditions (a system dependent on a potential, point attractors, gradient dynamics), it is possible to classify systems that are not stable in the sense of Whitney's theorem but whose evolution remains stable. Through this elaboration, we gain access for the first time to the movements of destabilization themselves, and we can distinguish situations (rare but central) in which destabilization follows a stable, hence universal, evolution. It can also be demonstrated that a dynamic system, stable in all its parameters except one, is controlled by this unstable parameter. Instability with stable evolution, therefore, governs stable development and makes the evolution of the system predictable. The science that, by definition, seeks the laws of system dynamics is justified in focusing on this class of strong and general processes (Wildgen 1999: 54–55).

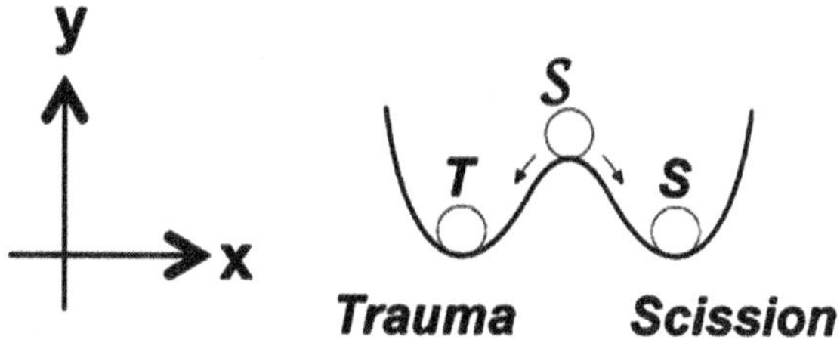

Fig. 7.2 Internal Space (x, y) of the cusp

What is then the centre of this repulsive field: the train station or the hospital? The answer lies in the morphogenesis of this district, which has seen the construction of no less than five successive dykes to prevent any permeability with the city.

7.2.1.1 The First Dike

(1) As seen in Fig. 7.1, the General Farmers, the former administrative limit of Paris, is surrounded by the current configuration of Paris. The primitive dike of *Boulevard de l'Hôpital* emphasizes the existence of a border that functions more like an "enclosure." This "enclosure" encircles the city of Paris from an administrative perspective on the one hand, and, on the other hand, separates the city from the women's prison of *Salpêtrière.*

7.2.1.2 Internal and External Space of a Cusp

In this article, we will show how this place (initially a prison, later a psychiatric hospital) is transformed in terms of its urban forms in Figs. 7.1, 7.5, 7.6, 7.7, and 7.8. Simultaneously, in the internal space of a wrinkle, we will depict the city when traversed by a world of Madness with its inherent TRAUMAS and also a world of living together with its SCISSIONS: TRAUMA (T) and SCISSION (S). In this case study, the *Salpêtrière* district has been structured at the rhythm of each new border (as we will progressively demonstrate), thereby presenting us with new modes of space[4] (b)and time[5] (a) organization, as shown in Fig. 7.2.

In Fig. 7.2, we present the internal space of the cusp with a first variable (x) and then a second variable (y). To better understand this attractor in the repulsion field that is the *Salpêtrière*, let's observe the external space of a cusp, one of Thom's catastrophes that allows us to understand the trajectory of two conflicting attractors, as seen in Fig. 7.3.

Figure 7.3 introduces, within Gilmore's framework (that we use to depict the morphogenesis of the *Salpêtrière*—Figs. 7.1, 7.5, 7.6, 7.7, and 7.8) a third variable,

[4] We use here the notion of space as a « relation of forces » (local and global) in conflict in an urban system.

[5] We use here the notion of time as historical time, the time of cultures, of societies which unfold and materialize regularly in a "state of morphologic stability".

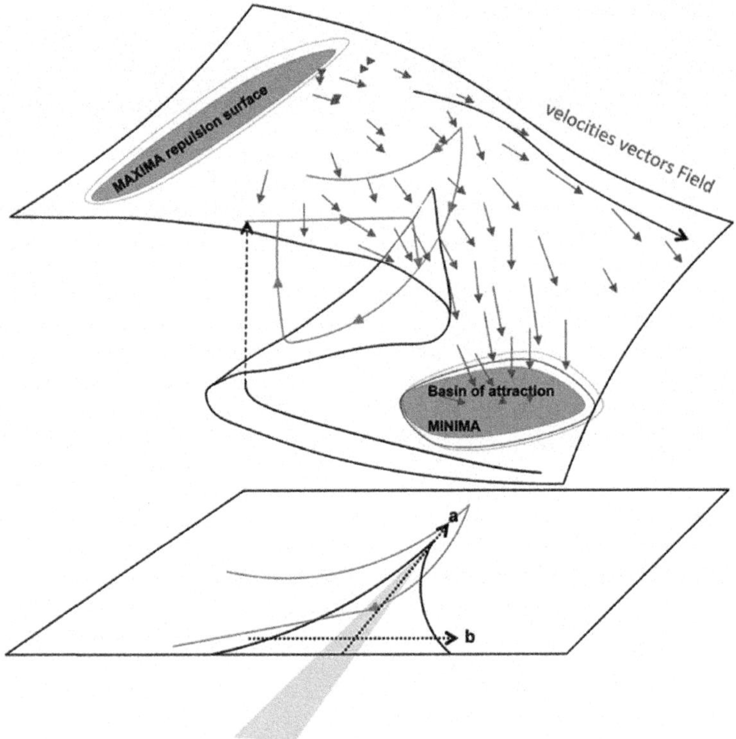

Fig. 7.3 We draw inspiration freely from the works of Brandt and Viret, who continued the work of René Thom, particularly around topological schematization

the gravitational potential (a) that, in our case study, corresponds to the organization of space as a "relation of forces"—both local and global—in conflict within an urban system. Additionally, a fourth variable in our case study corresponds to historical time, the time of cultures and societies that unfold and are regularly concretized in a "state of morphological stability" [29].

Variations in this potential field make possible an orientation of the surface with a "high" and a "low," between which there is a difference in level (potential), a difference in altitude (a), and longitude (b). We can then say that we have a repulsion field above and an attraction field below [9]. If we introduce into this cusp the internal variations of the system, we obtain Fig. 7.4, which allows us to understand the variations of the semantic field (T) Trauma (World of Madness)/(S) Scission (World of Living Together). In the mathematical and topological diagram [28], T and S are (x, y) attractors located in the conflict zone of control variables (a, b) of the potential (y = x4 + ax2 + bx), while the subjects, individual or collective, are the systems undergoing attraction.

In the perspective of Territorial Psychoanalysis, and referring to the diagram in Fig. 7.4, the city, like the human psyche, possesses two entities that can be considered

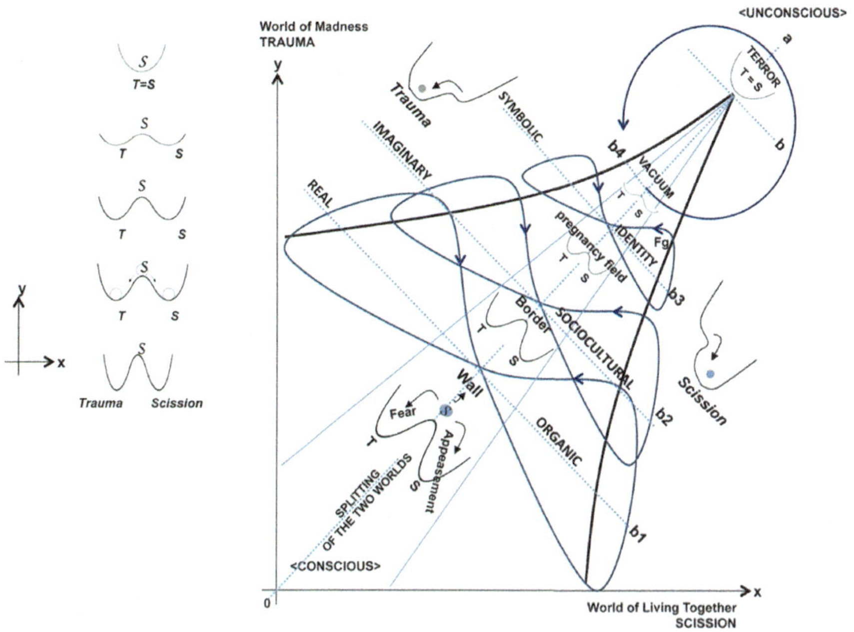

Fig. 7.4 summarizes the different phases of the morphogenesis of the Salpêtrière district, where this first dike marks the organic level (the Real instance, in accordance with Lacanian topology, see axis b1), where two internal spaces are invested by two passions, fear (trauma) and appeasement (scission)

as two attractors of the psychic variety: trauma (T) and egoic scission (S). The intensification of fear generated by the first attractor (T) towards the conscious self can lead the city, as a collective subject, or the individual, as a psychic subject, to feel the potentially terrifying threat of disorganization. The role of the territorial analyst is precisely to promote the egoic integration of an unconscious element, repressed and separated from the civic consciousness of the Urbis, in this case, the "city of fools." In this regard, the morphogenesis of the *Salpêtrière* illustrates the different phases of the historical process of egoic scission of an attractor-repulsive urban place, which is repressed from the collective consciousness. The intervention of the territorial analyst will primarily aim to assist the decision-makers of the urban development operation in resolving this historical scission of the Real (R), the Imaginary (I), and the Symbolic (S) instances in the continuity of the *polis* (see Fig. 7.4). It allows for the imaginary capture that enables the transformation of the *Vacuum* (*Salpêtrière*), an unthinkable and unsignifiable element (β element, Bion [6] instance R, Lacan 1967), into a thinkable and reintegrable α element [6] in an aesthetically reinvented urban Imaginary (I), which, finally, will lead to the ethical integration of the place (previously a repulsive attraction field) into the city where people live in the community. Following this clarification about the understanding of Fig. 7.4, we continue to expose each stage of morphogenesis.

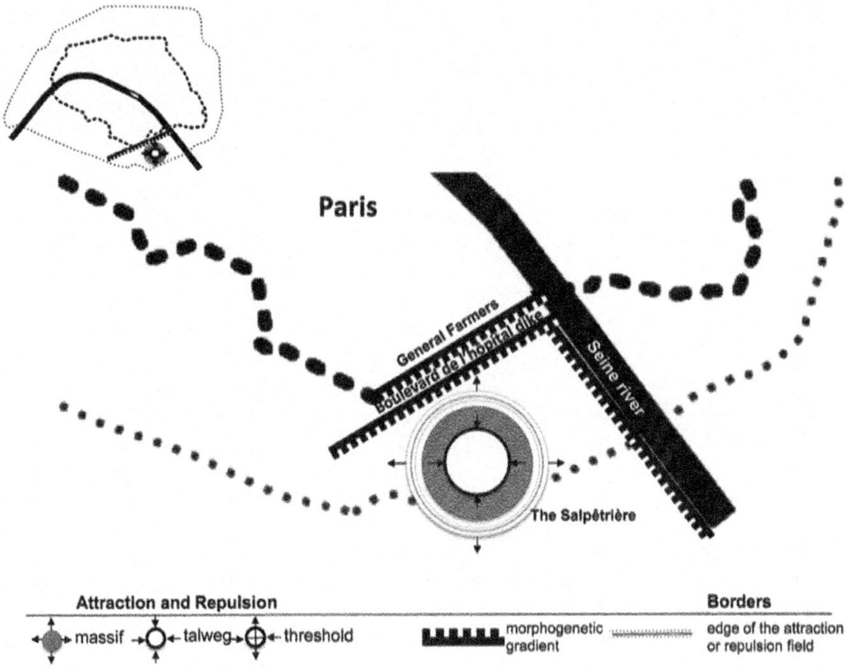

Fig. 7.5 The General Farmers dike sanctions the ban on building beyond this limit from 1786

7.2.1.3 The Second Dike

(2) The dike represented in Fig. 7.5 transformed the *Salpêtrière* from 1786 on into a place of recuperation for what the city rejects (beggars, prostitutes, the mentally ill, etc.). Consequently, this location became an "attractor" for activities rejected by the city and a "field of repulsion" for the city. The hospital is separated from the city by a double barrier that establishes the impossibility of staying there unless one belongs to a group of rejected people (see Fig. 7.3).

In Fig. 7.4, we depict a new phase of morphogenesis, situated at the sociocultural level—Imaginary (I)—in accordance with the Lacanian topology, see axis b2. The prohibition of building in this area is marked by two dikes: those of the General Farmers and the River Seine. Two centuries after the construction of the hospital and the first dike this space splits into two spaces marked by Trauma and Scission in their relationship with the city of Paris.

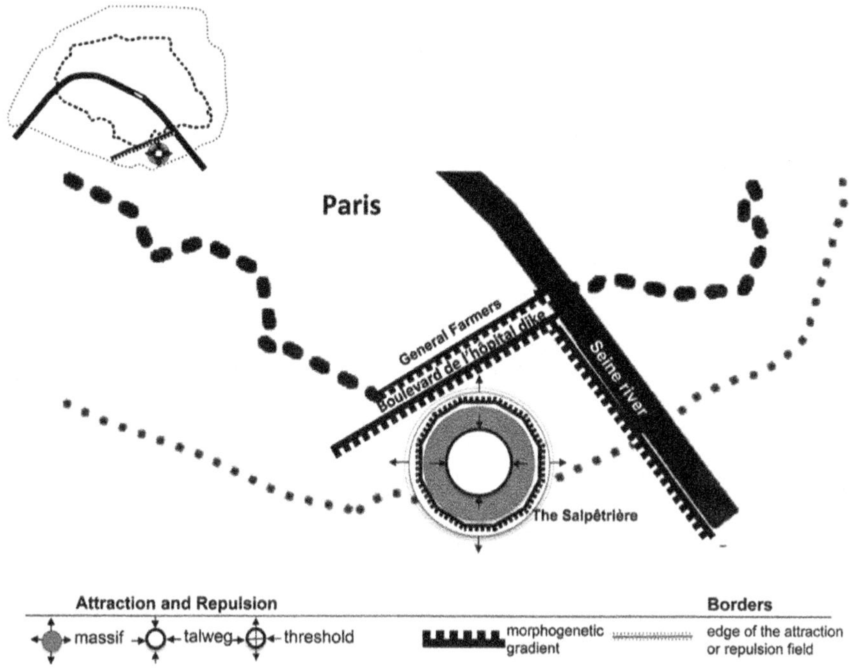

Fig. 7.6 The dike is reinforced by the institutionalization of the psychiatric hospital in the 19th century and becomes a pregnantial field

7.2.1.4 The Third Dike

(3) The dike represented in Fig. 7.6 in the 19th century became a pregnantial field. For centuries, the *Salpêtrière* has acted as an "attractor in a field of repulsion" located "at a good distance" from the city and "prohibited" from urban development.

In Fig. 7.4, we depict a new phase of morphogenesis, situated at the level of collective identity (equivalent to the Symbolic instance, see axis b3), as psychiatric institutionalization symbolically reinforces the traumatic space and definitively separates it from the city of Paris. The advantage of this dynamic model is that it shows both opposition and interaction, even the possibility of continuity and fusion. The four strata (b) are the horizontal axes separated (dotted lines) within the vertical (a), representing the bifurcation into two worlds (the world of madness and the world in which we live together, in the city): without the existence of these opposed external worlds, represented in the cusp diagram, there would be neither structure nor structuring of the individual (social subject) and the person (psychic subject).

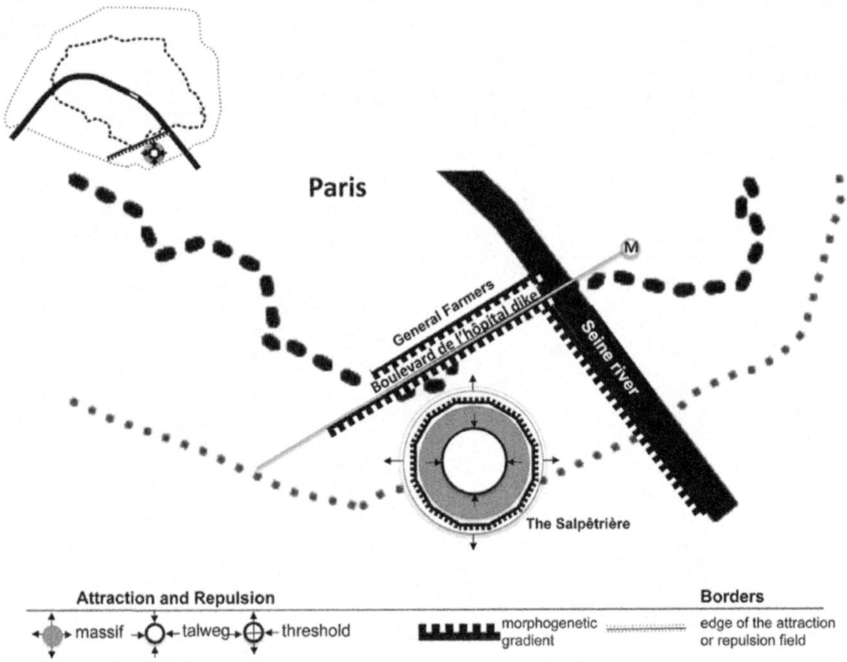

Fig. 7.7 The dikes: in 1840

7.2.1.5 The Fourth Dike

(4) The Dikes:

- The first physical dike, in 1840, materialized by the construction of the railway line, isolating the hospital, and transforming it into an urban island.
- The second dike, functional in 1906, is the viaduct of the aerial metro, blocking the perspective of the hospital's monumental facade.
- Successive dikes highlight the existence of several morphogenetic gradients (see Fig. 7.7).

7.2.1.6 The Fifth Dike

(5) Patrimonial dikes, with the classification of the chapel, the main façade of the hospital (see Fig. 7.8), and the site of Place Saint Louis. Another patrimonial dike appeared in 1997 with the classification of the railway station as a historical monument. The conjunction of all the dikes described so far reveals morphogenetic gradients that destabilize the urban fabric, a destabilization centred around the hospital (even if its psychiatric function is no longer active). Figure 7.4 cusp shows that when the morphogenesis of the *Salpêtrière*

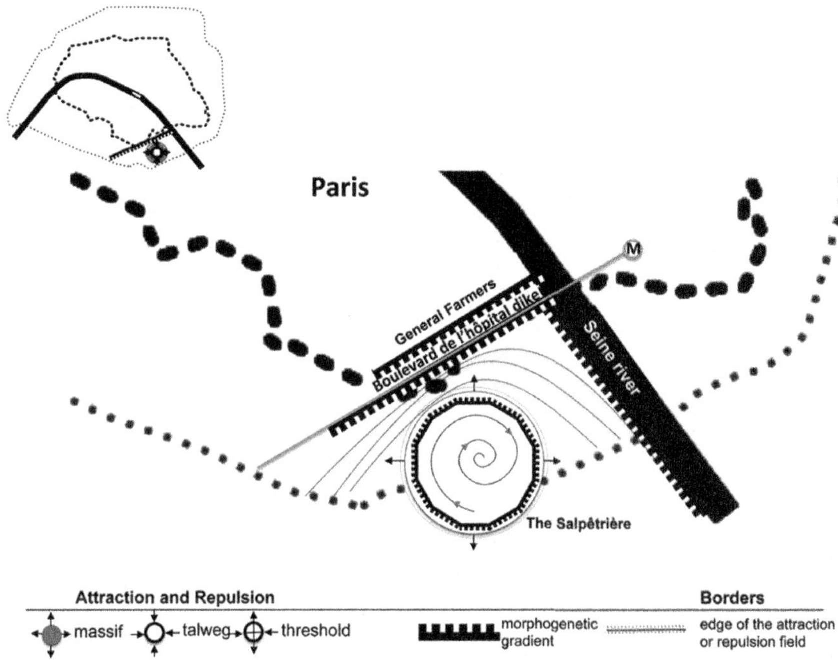

Fig. 7.8 The patrimonial dikes appeared in 1976

is saturated, axis b4 becomes a spiral that starts turning on itself, resulting in T = S (on the cusp tip).

How can we analogically derive this morphogenesis of the *Salpêtrière*, an attraction centre in the middle of a repulsion field of the physical body of a (no) habitable place, through a system of dikes transformed into true morphogenetic gradients that destroy the histology of the urban body, towards the psychic body of a human subject in existential suffering? The attempt to answer this question follows.

7.2.2 The Woman Pitié: Regarding Individual (Psychic) Space

Pitié is a 38-year-old woman with a clear skin tone and blue eyes, of medium height and delicate build, with an elegant silhouette and a captivating demeanour. As an environmental systems engineer, she seems to have achieved professional success, holding a doctorate from Stanford University.

Her marriage to Martin, a neurologist thirteen years her senior, is marked by unhappiness, and she is the mother of a fourteen-year-old son, Lucien, whom she deeply cherishes. Pitié comes from an upper-middle-class family with a high intellectual status. The eldest of three sisters, admired and adored by an idealized father, a renowned physicist and former university physics professor. Her relationship with her mother is somewhat ambivalent, characterized by weak homosexual identification and superficial female complicity. Pitié feels that her mother has taken a back seat compared to her father. Conversely, she has always shared a profound interest in natural sciences, geology, wildlife, and science fiction with her father. Together, they used to embark on long countryside walks, engaging in discussions about plants, rocks, animals, and watercourses, while her sisters stayed at home with their mother, playing and watching television. At the age of 10, still a prepubescent child, Pitié underwent a traumatic event that remained non-memorable for an extended period. Her maternal uncle, Joseph, a doctor at the *Salpêtrière*, intelligent and seductive, whom she dearly loved as a child (he often brought her to the *Salpêtrière* to play doctor), sexually assaulted her in an obscene and intrusive manner during one of her playful visits to the hospital. As a preadolescent, it became unbearable for her to visit the *Salpêtrière*. Following this sexually aggressive act, of a traumatic intensity correlative to her tenderness for her uncle [1, 14], she wasn't well. Around puberty, by the age of 12, her uncle, despite seemingly denying the intrusive seductive act, advised her parents to take her to one of his colleagues, a psychiatrist and psychoanalyst. The preadolescent underwent a short and somewhat problematic therapeutic process. In fact, whenever Pitié entered this traumatic space, the traumatic "après-coup" returned in the form of a collapsing episode, as if this place carried an unmemorable unconscious signifier, an alarming and unbearable "sinthome" [25] stemming from a Real instance split from an elusive Imaginary, with lack of Symbolic mediation. Eventually, her parents decided not to take her to this doctor anymore. This "thing" (*Das ding*, [24]), this "semiotic unknown" [32], omitted from consciousness and repressed as implicit memory, in the realm of hysterical suffering, went through a dramatic transformation during her youth—from an attractive and carefree young girl to a shy and reserved teenager. Pitié actively avoided any encounters with her male colleagues, even though this prospect often excited her libido. She was eventually diagnosed with irritable bowel syndrome, further restricting her social life. At the end of her studies, she finally started dating a colleague of the same age, a quiet, polite, and timid man named Jean. Their sexual life remained sporadic and unpleasant. She did not accept preliminary masturbatory caresses, and Jean suffered from premature ejaculation. However, during this time, irritable bowel syndrome entered a (almost) spontaneous remission and was replaced by anorgasmia. We can consider that irritable bowel syndrome and anorgasmia are hysterical-dissociative symptoms that function as an inhibitory "dyke" to her adult sexuality, evoking a "morphogenetic gradient" that, like one of the "dykes" at the *Salpêtrière*, threatens to destroy her femininity. Indeed, it is a "dyke" that isolates, like a morphogenetic gradient, Pitié's erotic-sexual drives, which constitute an uninhabitable place repoussé/split from the rest of her personality. Ultimately, the sexual inhibition of Pitié was partially and temporarily "rescued" by the birth of her son Lucien and the maternal care she provided him.

On Lucien's third birthday, she ended her relationship with Jean, and four years later she met Martin, a neurologist thirteen years her senior. Despite being attracted to this strong, successful, seductive, determined, and mature man, she couldn't help feeling defensively anxious, almost threatened in her intimacy by their physical proximity. Their sexual intimacy resumed to copulation, as Pitié avoided caresses or other erotic-masturbatory preliminaries. Martin eventually accepted what he attributed to his partner's shy temperament and began to moderate his amorous and sexual impulses. Six months after their marriage, almost seven years ago, they went on a weekend getaway to a charming hotel in a picturesque coastal town. After dinner and a drink at the bar, they returned to their room. She was tired from the day's walk, while Martin was sexually aroused. He attempted to make love to her, but she gently postponed his desire to the next day. After sleeping for some time, she woke up abruptly, in a state of shock, feeling Martin's fingers rhythmically caressing her vulva while erotically fondling her breasts with the other hand. Shocked and trembling, she jumped out of bed and, in a panic state, quickly dressed and left the room without saying a word, silently crying. Martin tried to follow her, but she dissuaded him with a nervous gesture.

She recalls walking on the beach all night until sunrise. Without being able to think, dominated by an invasive "beta screen" [6], Pitié acted in the form of an "après-coup" the "sinthome" in this place-hotel that phantasmatically became the physical, also the physico-symbolic, original traumatogenic place, her uncle's consulting room at *Salpêtrière*. When she returned to the room in the early morning, after the phantasmatic revival of her past-present "traumatic memory" she found Martin insomniac and tense. She kissed him on the cheek, took a bath, got dressed, and they went down together for breakfast without exchanging a word about what had happened. It was as if this invasive traumatic "after-coup" had deepened the severe egoic split already at work, through the psychic operation of the "atomization" of her personality, in the sense of [15], further accentuating the dissociative phenomena that would take place later.

Indeed, it was after this traumatic episode that Pitié urgently scheduled an appointment with her psychiatrist-psychoanalyst, who, in such urgency, could only meet her at his cabinet at the *Salpêtrière* Hospital. She tried to make her way to the hospital/ hallucinatory "sinthome," but could not find the way to this traumatogenic place. She felt stunned and lacking psychomotor coordination during this dissociative phenomenon, which acted as a kind of psychic "dyke" that unconsciously made her flee the strange attractor of a traumatic repulsion field, and automatically head in the direction of the cabinet where her psychoanalyst, always psychically dissociated from the psychiatrist at the *Salpêtrière*, sees her in psychoanalytic session three times a week, in a comfortable cabinet located in a peaceful district of the Parisian capital, the 15th, which she appreciates for the comfort and serenity she feels there all the time (like the comfort of her parents' vacation home in her childhood).

7.3 The *Vacuum*

The *vacuum* is a highly significant place for all human subjects because it is invested with symbolic meanings that are both attractive and repulsive. According to Ritchot ([33]: 120), it is an empty box in the geographic area that plays a major structuring role in the morphogenesis of the city. In this section, we will explore the way the morphogenetic development of this neighbourhood led us to the discovery of a *vacuum*-type organization (i.e., of a space of attraction and repulsion) [30], and to the way the awareness of intervening in a *vacuum*-type space helps to solve the symptom of an unresolved historicity disconnected from the urban fabric surrounding the *Salpêtrière* district.

In the case of Pitié's life, the *vacuum* is represented by an Ego Ideal split from the (repressed) incestuous sexual drive directed towards the Oedipal Father. It is also represented by the maternal-feminine investment in a child-couple family configuration, which, after her personal analysis, partially succeeded the secondary identification either with the Oedipal Mother or with the Oedipal parental couple. Let's observe how each example, one after the other, allows us to understand the connection between the traumatic organization of the *Salpêtrière* district and Pitié's internal place of trauma.

7.3.1 The Example of the Salpêtrière

The urban fabric is composed of a hospital monument from the 17th century, historically marked by "rejection." The main façade of the hospital was built to give a certain dignity to its function of concealing what the city rejects (beggars, prostitutes, madmen, etc.). In this context, this façade symbolically imposes itself as a "rejection" wall and as the mark of an introverted island/of an old prison. The metabolic function of this space acts as a structural "enclave" that brings around it the "rejection" and operates as an attractor pole of "rejection" (see Fig. 7.7). Parisians feel repelled by this place, a kind of bad *genius loci*. The "spirit of the place" is formed by the juxtaposition of a historical 17th-century hospital, a historical 19th-century train station, and an interstitial space with indistinct and confused borders [30]. This "residual" space is the visible expression of the effects that the "repulsive attractor" exerts on the surrounding space, as observed in Fig. 7.7. It functions as a *vacuum* that generates a dysphoric space around it and spreads inside the train station, invaded by disparate activities unrelated to the place. Only a reorganization of this devalued urban mesh will be able to influence the symbolic values of the place, allowing it to be valuably invested (to be valued) and to operate a symbolic shift towards another "spirit of the place" (see Fig. 7.9).

The urban proposal to create a space by acknowledging the presence of a *vacuum* (a space of attraction and repulsion) within the hospital's space helps to solve the

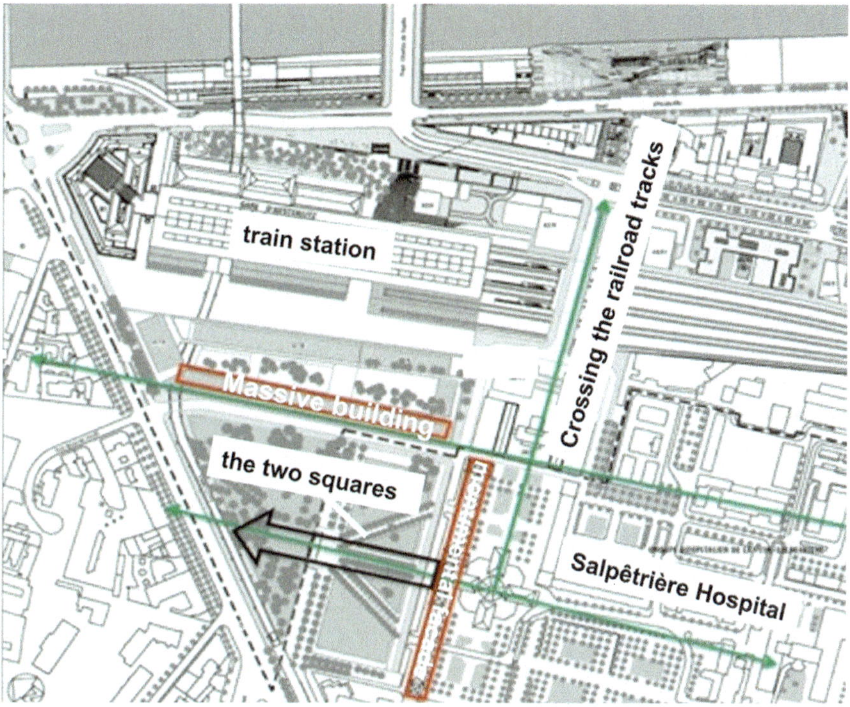

Fig. 7.9 Map of the new Salpêtrière district. [30] *Source* SEMAPA and Marcos and Bayle

symptom of an unresolved historicity. As observed in Fig. 7.9, the site of the church with patrimonial interest is used to symbolically reinvest this place through:

1. The establishing of axes, indicated in green in Fig. 7.9, that open the hospital space to the city.
2. The crossing of the railroad tracks (emptying that way one of the morphogenetic gradients).
3. The redesign of the two squares into one in front of the monumental facade of the hospital. See the black arrow representing the union of the two squares and the link that can now be established with the city.
4. The construction of a massive building perpendicular to the main facade of the hospital, allowing:
   - The creation of a border that unites by separating and separates by uniting the train station and the hospital.
   - To clearly mark the boundaries of the new square.
   - To interweave spaces that were previously disconnected.

Each of these elements will gradually integrate the *Salpêtrière* district into the urban fabric by transforming the dikes into places valued by the passage of residents and foreigners.

7.3.2 The Example of Pitié

Regarding Pitié's psychic space, the reconfiguration of the *vacuum* from adult sexuality that is both attractive and repulsive, also resulting from an unresolved subjective historicity, involves overcoming the morphogenetic gradient inhibiting her sexuality. In the dysfunctional unconscious configuration, the *Id* (the drive pole in Freudian structural topography, 1923/1969) remains rejected/split from consciousness (*ego*—reudian structural topography), or, according to the Klein-Bion perspective [6, 22], the β protomental-dream elements, still unthinkable, reinforce the "fight-flight" basic assumption [3] in relation to her sexual instincts. This primitive and unthinkable emotional *sensorium* can also be considered as "the thing" (*Das Ding*, [24]) that stimulates hallucinatory jouissance "beyond the pleasure principle" [17], either through cumulative trauma episodes [21] described above or through the following traumatic dream brought by Pitié to an analytic session: she finds herself in front of a beautiful country house, which she identifies as her parent's vacation home. It's a sunny day, and she's in a good mood. As she enters the house the scenario changes drastically. The house seems somewhat different, strange and resembles more to the inside of a hospital ward. To her surprise, the space is empty, as there is no one inside. She glimpses what appears to be a medical cabinet at the end of a corridor surrounded on both sides by empty beds. The space seems hybrid, but she retains the impression of a large infirmary-like room immersed in a hazy and surreal atmosphere. As she gets near the medical cabinet, she realizes it is smaller than it seemed from the outside. She reaches the cabinet's entrance door, and the air around her becomes heavier. Upon entering, she realizes the furniture is covered with white sheets. The atmosphere is dusty, she struggles to breathe, and she feels almost suffocating. She wants to leave the cabinet but lacks the strength and feels terrified. As she wanders in this phantasmagorical scenario, she notices the furniture is moving and trembling strangely, and a mysterious and frightening physical sensation seizes her body. She wakes up in a panic and continues to feel this strange hallucinatory sensation, this "hallucinatory excretion" [5] throughout her body. It is important to notice that the dream, like the traumatic-hallucinatory episodes, acts as a "repulsive attractor," whose function is to keep Pitié in "quarantine" regarding the bodily, instinctual-drive dimension of her adult life. However, the scenario of the traumatic dream seems to confirm the phantasmatic collusion of the sexual body, which is the somato-psychic dimension of her fragmentary neurotic distress, and the physical space of the hospital (especially the "cabinet" where she feels captive and seems to be the traumatic reminiscence of her uncle's medical cabinet at the *Salpêtrière*, the place of the original traumatic experience). Furthermore, it's as if the unconscious language of the "parlêtre" [23], which is the effect of the Symbolic on the Real of a mysterious body [34], could be taken up in the imagistic language of the dream, as a Symbolic instance making the Real of an enigmatic physical space "speak" via a "parlespace" that has the value of the linguistic dimension of the phantasmatic collusion of the subject's sensual-instinctual body and the physical space surrounding the subject.

Let's revisit the cusp in Fig. 7.4 (see Sect. 7.2.1.2 of this chapter) to further understand the various affective states in the aftermath of the original traumatic situation (see Fig. 7.11). The material of this traumatic dream can be represented by a humoral topology of the cusp in Fig. 7.10, where the "euphoric" and pleasant side—sunny day, house façade, and good mood—is transformed into the "neutral" intermediate space of psychic asthenia, alternating between anxious impassivity, with a "morose" note [27]—the large infirmary-like room immersed in a hazy and surreal atmosphere—and delusional hallucinosis—the phantasmatic scenario where the furniture moves and trembles strangely, and the mysterious and "frightening" physical sensation that seizes the dreamer's body. Finally, the "dysphoric" side appears when transitioning to the waking state, at the dawn of consciousness, when Pitié wakes up in a panic—fear—gripped by the terrifying sensation of continuing to feel the strange hallucinatory sensation invading her entire body.

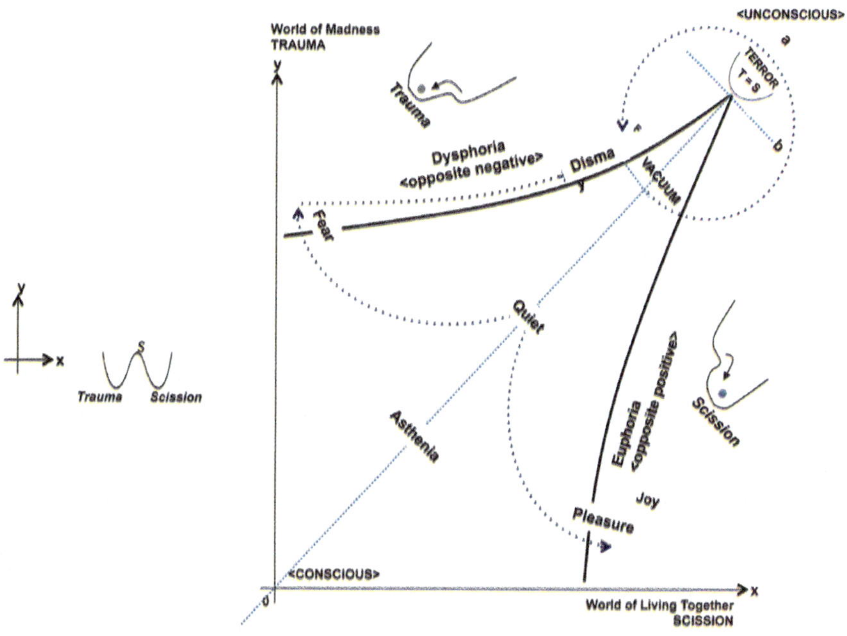

Fig. 7.10 Humoral Topology (Brandt and Marcos) to understand various thymic states in the aftermath of the original traumatic situation

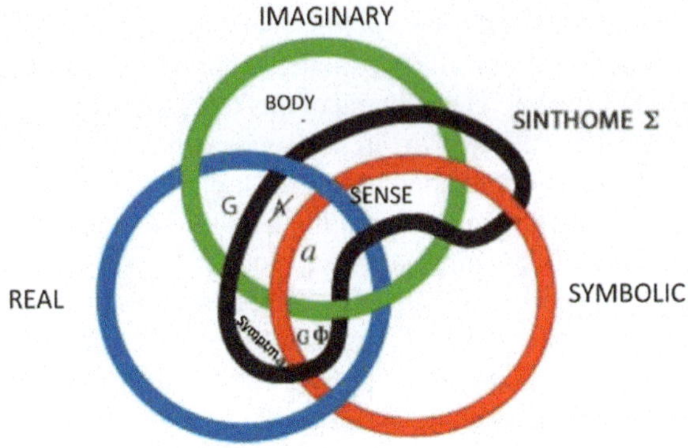

Fig. 7.11 Borromean Knot consisting of 4 interwoven elements corresponding to the 4 registers of subjectivity. *Source* Farate, C., Santos Jorge, M. (Eds.) (2018) Bion e Lacan para além de Freud: ideias, signos e "palavras-cruzadas" para uma psicanálise em transformação. Almedina

7.4 Symptom—Morphogenetic Gradient

7.4.1 The Example of the Salpêtrière

The morphogenetic gradient can be compared to a border front that oscillates between the two extremes of this gradient. It can also be defined as a field that moves or gradually crystallizes. In the example of the *Salpêtrière* district, the morphogenesis allows us to identify an accumulation of dikes, dams, borders, border fronts, fields, pregnancies, and morphogenetic gradients that are symptomatic of a "repulsive attractor" of the former hospital of the insane (see Figs. 7.1, 7.5, 7.6 and 7.7). In other words, the *Salpêtrière* district, according to the concepts of Desmarais (1995: 93–95) and Marcos [28], has become a massif that can no longer develop, becoming an enclave in an urban space. The presence of this enclave hinders the deployment of morphogenetic gradients, forcing certain trajectories to bypass this district. The *Salpêtrière* Hospital has become over time a *vacuum* as it is placed between two concentric borders.

7.4.2 The Example of Pitié

Pitié's prepubescent sexual desires towards the idealized Oedipal father, displaced towards libidinal interests in exciting earthly objects (stones, streams, plants, soil), were fantasized and revealed as a seduction trauma by Joseph, an erotically invested paternal substitute, during unusual "anatomical lessons" symbolically marked by innocent play in the hospital's cabinet. In these "lessons," Pitié experienced a strange pleasure sensation through masturbatory stimulation of her *sensorium* until she felt dramatically invaded by a shocking and abusive genital stimulation. Somatic-functional and hysteric-dissociative symptoms—shyness, social withdrawal, irritable bowel syndrome, anorgasmia—acted not only as a "morphogenetic gradient", or a specific "pregnantial field", through the dissociation between sexual impulses and erotic fantasies and the normative intellectual interests and investments. They also acted as a "sinthome," in Lacan's sense [25]. Thus, the Real instance (R), the biological-sensual unknown, disconnected from the Imaginary instance (I), could act, through the "sinthome," as a corporeal and relational "dyke-dam" still able to connect to the Symbolic instance (S), though in a split and dissociative, yet non-psychotic, manner (see Figs. 7.8 and 7.11).

7.5 Border

7.5.1 The Example of the Salpêtrière

In what relates to the apprehension and reconstruction of the form of the *Salpêtrière* district through morphogenesis, the "bordering"[6] is operated through the following sequence (see Fig. 7.4):

1. "Enclosure" of the hospital boulevard (organic level: Real instance—axis b1).
2. "Border" of the General Farmers (socio-cultural level: Imaginary instance—axis b2).
3. Institutional transformation into a psychiatric hospital in the 19th century, giving rise to a "pregnantial field" (level of collective identity: Symbolic instance—axis b3).
4. Multiplication of borders—physical, functional, patrimonial—giving rise to several "morphogenetic gradients" surrounding a *Vacuum* around the hospital, destabilizing the urban fabric. The morphogenesis is saturated, and axis b4 becomes a spiral that starts turning on itself: T = S.
5. Symbolic investment of the hospital church space, generating two diagonal axes at the level of the former pregnantial fields, allowing for the gradual integration of the *Salpêtrière* district into the urban fabric.

7.5.2 The Example of Pitié

The border lines that maintain internal and external limits in Pitié's life, despite the *vacuum*/repulsive attractor/schizoid-paranoid field (vertex PS, [7]) that undermines her sexual freedom as a woman and jeopardizes her narcissistic-objectal equilibrium, emerge from the previously mentioned *vacuum*. These border lines operate

[6] The use of the concept of "bordering," and not "border," presupposes at its root the presence of a hierarchized dynamism, staged through several morphological levels of organization of the urban psycho-sphere [31]. These levels—Real, Imaginary, and Symbolic—allow us to grasp the morphogenesis of the *Salpêtrière* district in all its complexity.

in a dual way: the Superego- Ego Ideal remains split from primitive, hysteric-dissociative, and traumatic-hallucinatory (β elements), allowing Pitié to invest in an intellectual-professional career; the "contact-barrier-α" operation [4] authorizes the non-psychotic investment of an adult mother–child-couple boundary line, through the elaboration of the depressive position (further reinforced by her personal analytic work).

7.6 Pregnancy—Pregnantial Field

7.6.1 The Example of the Salpêtrière

Considering the theory of "saliences" and "pregnancies" introduced by René Thom, we can specify that a pregnancy or pregnantial field can be understood, for example, as certain forms that have (in animals) biological significance, such as prey for the predator (hungry), or the predator for the prey, the sexual partner at the appropriate time. The recognition of these forms elicits a strong reaction in the subject: release of hormones, emotional excitement, attraction or repulsion behavior towards the inducing form, which we can call pregnantial forms. In this perspective, the *physical substrate* of the city of Paris appears as a fluid, a highly pregnant background from which salient forms detach, where the space is differentiated by qualitative discontinuities: "the ground of our experience" ([28]: 149). The notion of pregnancy allows for the development of a concept that connects subjects to their valued objects. It emphasizes that the grasp of meaning is a deep process of subjectivization of semantic pregnancies that can only occur through an actantial syntax, i.e., through value investments in objects. These are "instinctual" and "unconscious" contents, both interoceptive and proprioceptive, therefore primarily thymic in nature. Thus, the *Salpêtrière* district initially appears as a semantic pregnancy. As the virtual contents are gradually actualized by an actantial syntax, as we observed in the deployment of its morphogenesis, these pregnancies turn into "pregnantial fields" and transform the hospital space into a "repulsive attractor." This repulsion occurs when objects of value (in this case, the hospital space) are invested with attributes perceived as negative by the subjects traversing this space.

7.6.2 The Example of Pitié

Pitié's pregnancies are based on internal and external perceptions that structure her morphogenesis. There is clearly a sensualist pregnantial field during her childhood and adolescence that shaped her feminine identity:

1. The libidinal investment and secondary identification to an idealized Oedipal paternal figure.

2. Conversely, the highly guilt-inducing incestuous-seductive traumatic fantasy has repressed/pushed away her feminine physicality until now, while significantly restricting her sexual-loving investments.

This dual pregnantial field, interpretable psychoanalytically as a "common psychic field" for Pitié and the Other [13], in this case, her psychoanalyst, still dissociated/split from the psychiatrist-psychoanalyst she consulted at the *Salpêtrière*, was introjected as a "fragmented" libidinal body [15, 20], as a traumatogenic physical space, which opens the way to the deepening of the conflict dynamics within the framework of her personal analysis and will gradually make possible the ultimate resolution of her sexual inhibition.

7.7 Salience

7.7.1 The Example of the Salpêtrière

René Thom's theory of pregnancies and saliences emphasizes the substantial and affective content of values. In the pregnantial field surrounding the *Salpêtrière* district, a salient form stands out: the hospital chapel. It is salient because it plays a structuring role in the *Salpêtrière* district. It is a form that is emptied of its initial role (place of worship), and reinvested with urban values, allowing for the gradual integration of the *Salpêtrière* district into the urban fabric while transforming the dikes into places valued by the passage of residents.

7.7.2 The Example of Pitié

The pregnantial field defined by the unconscious collusion of the female body and the physical "body" of the hospital generates a dual salience (action-behavior): (1) on the one hand, the choice of a maternal-marital objectal investment and personal analysis, on a cabinet/non-egoic setting [8] reassuringly away from the hospital, is likely to free her from adult sexual inhibition in the future; (2) on the other hand, the ongoing confinement in a hysteric-dissociative symptom (vaginismus) and the inhibition of sexual intimacy and of a non-saturated representation of the traumatic memory of the physical space around the *Salpêtrière* still jeopardizes both her couple relationship and her freedom of displacement in places that continue to be, consciously or unconsciously, "targeted" by the saturated traumatogenic memory.

7.8 Concluding Remarks

This chapter whose aim is to introduce a new and innovative interdisciplinary domain—Territorial Psychoanalysis—is structured around two case study descriptions. The first case focuses on the *Salpêtrière*, an urban-institutional collective space historically significant for psychoanalysis, psychiatry, and the city of Paris. The second case, an individual case study, refers to the clinical vignette of a woman named Pitié, who has been in personal analysis for three years.

As we described the socio-urban case (*Salpêtrière*) and the individual case (Pitié), the complexity of factors summoned, from the side of the urban space, internal and external dimensions, and on the side of the human psyche, the dimensions of the person's relationship to the world. As a result, we used several models from the theory of the catastrophes reared and reinterpreted by Brandt [10–11] not only to deepen the interest in this new empirical-conceptual domain but also to make it more intelligible to our readers:

1. First, we used Gilmore because his model allowed us to represent a field of attraction or repulsion with several separatrices, crest lines, thalweg lines, and passages. This study of urban morphogenesis allowed us to discover a latent unconscious field in the organization of the *Salpêtrière* district. We studied it with the help of another model: the cusp.
2. Secondly, the introduction of the collective unconscious as an operative element in the organization of space led us to present the internal space of a cusp. Thus, we represented the city as much when it is traversed by a world of Madness with its inherent Traumas (T), as when it defines itself as a world of Living together with its inevitable Scissions (S). The *Salpêtrière* district has structured itself, in this case study, at the rhythm of each new border (as we have shown progressively), and thus, this district confronted us with new modes of space (b) and time (a) organization, as we can see in Fig. 7.2. In Fig. 7.2, we presented the internal space of the cusp with a first variable (x: World of Madness) and then a second variable (y: World of Living together).
3. Thirdly, to better understand the external functioning of an attraction and repulsion field like the *Salpêtrière*, we were freely inspired by the work of Brandt and Viret, who continued the work of René Thom, namely around the topological schematization itself. In Fig. 7.3, we introduced the Gilmore scheme (which we use to represent the morphogenesis of the *Salpêtrière*—Fig. 7.1, 7.5, 7.6, 7.7, and 7.8) by adding a third variable, the gravitational potential (of altitude (a) and longitude (b)).
4. Fourthly, we introduced the internal variations of the system into this cusp. In this regard, Fig. 7.4 allows us to understand the variations of the semantic field (T) Trauma (World of Madness)/(S) Scission (World of Living together). In the mathematical and topological diagram (Brandt and [28]), T and S are (x, y) attractors located in the conflict zone of control variables (a, b) of the potential ($y = x4 + ax2 + bx$), while individual or collective subjects are the systems that undergo the attraction.

5. Fifthly, Fig. 7.4 summarizes the different phases of the morphogenesis of the *Salpêtrière* district, separating two internal spaces invested by various passions as the morphogenesis reaches its saturation point. Axes b1, b2, and b3 articulate the two passions: fear and appeasement, which increase in intensity as we approach the cusp point. Axis b4 becomes a spiral that starts to turn on itself, implying that T = S moves with terror.

The articulation of various models from catastrophe theory and the theory of pregnancies and saliences has taken us on a unique research journey. We started with the morphogenesis / morphogenetic gradient of the *Salpêtrière* and Pitié, both oriented toward the current-past and past-current traumatic events, which shaped their historical and existential trajectories. The symptom has been schematically exposed and has evolved making obligatory reference to the concept of the *vacuum* not only to represent the internal 'fight-flight' towards the threat of the 'semiotic unknown' but also to account for the internal dynamics of the conflict that ultimately built the border in both cases. The 'bordering' constituted the key moment that symbolically filled the *vacuum* and allowed the semantic pregnancy/pregnantial field of the instincts that move along the 'dykes-dams' and define the space (cyclically) renewed to be perceived and thought by the subjectifying object and the objectifying subject. This journey has highlighted the importance of the fundamental concepts that build this new interdisciplinary field: morphogenesis, morphogenetic gradient, vacuum, symptom, border, pregnancy, and salience. The morphodynamic analysis method of urban construction, based on the intersubjective dialogical relationship between the inhabited human territory and the human subject inhabiting it, has progressively unveiled the foundations of this new and innovative interdisciplinary scientific field: Territorial Psychoanalysis.

This chapter is, thus, the first step in a long journey of interdisciplinary research that will allow us to understand more about the theoretical and operative undertaking of this essential and inextricable intersubjective link between the subject who invests, who builds, who shapes his environment and the environment that, in turn, constantly influences the subject who inhabits and invests it as a place of (hopefully) stable life.

References

1. Balint, M.: Trauma and object relationship. Int. J. Psychoanal. Psychoanal. **50**, 429–436 (1969)
2. Benjamin, W.: Le livre des passages (trad.fr. Jean Lacoste), Paris, Éditions du Cerf (1989)
3. Bion, W.: Experiences in Groups. Tavistock, London (1961)
4. Bion, W.: Transformations: Change from Learning to Growth. Heinemann, London (1965)
5. Bion, W.: Cogitations. Karnac, London (1992)
6. Bion, W.: Learning from Experience, trad. française, F. Robert, Aux sources de l'expérience. PUF, Paris 1979 (1962)
7. Bion, W.: Elements of Psycho-Analysis, trad. française, F. Robert, Éléments de Psychanalyse. PUF, Paris 1979 (1963)
8. Bleger, J.: Psycho-analysis of the psychoanalytic frame. Int. J. Psychoanal. Psychoanal. **48**, 511–519 (1967)

9. Brandt, P.A.: La charpente modale du sens. Aarhus UniversityPress, Pour une sémio-linguistique morphogénétique et dynamique, Aarhus (1992)
10. Brandt, P.A.: Morphologies of Meaning. Aarhus University Press, Aarhus (1995)
11. Brandt, P.A.: Spaces, Domains, and Meaning. Essays in Cognitive Semiotics, n°4, European Semiotics. Peter Lang, Bern (2004)
12. Danon-Boileau, L.: L'après-coup: devenir miraculeux du trauma ou coup d'épée dans l'eau? Revue Française de Psychanalyse **70**(3), 727–735 (2006)
13. Farate, C.: Psicanálise com Limites e Psicanálise Ilimitada: o setting como instrumento técnico e estrutura dinâmica da psicanálise. Fenda, Lisboa (2012)
14. Ferenczi, S.: Confusion of the tongues between the adults and the child–(the language of tenderness and passion). Int. J. Psychoanal. Psychoanal. **30**, 225–230 (1949)
15. Ferenczi, S.: Final contributions to the problems and methods of psycho-analysis. In: Balint, M. (ed.) Hogarth, London 1955 (1932)
16. Freud, S.: Além do Princípio do Prazer. In: Freud, S. (ed.) Obras Completas, vol. XVIII, pp. 17–75. Imago, Rio de Janeiro 1969 (1920)
17. Freud, S.: O Ego e o Id. In: Freud, S. (ed.) Obras Completas, vol. XIX, pp. 15–77. Imago, Rio de Janeiro 1969 (1923)
18. Gilmore, R.: Catastrophe Theory for Scientists and Engineers. Dover Publications, New York (1981)
19. Green, A.: La folie privée–Psychanalyse des cas-limites. Gallimard, Paris (1990)
20. Gutierrez Peláez, M.: Trauma theory in Sandór Ferenczi's writings of 1931 and 1932. Int. J. Psychoanal. Psychoanal. **90**, 1217–1233 (2009)
21. Khan, M.R.: The concept of cumulative trauma. Psychoanal. Study Child. Study Child. **18**(1), 286–306 (1963)
22. Klein, M.: Notes on some schizoid mechanisms. Int. J. Psychoanal. Psychoanal. **27**, 99–110 (1946)
23. Lacan, J.: Le Séminaire, Livre XX, Encore. Seuil, Paris (1975)
24. Lacan, J.: Le Séminaire vii : Éthique de la Psychanalyse. Seuil, Paris 1986 (1959–60)
25. Lacan, J.: Le Séminaire xxiii: Le Sinthome. Seuil, Paris 2005 (1975–76)
26. Largeault, J.: Principes de philosophie réaliste. Klincksieck, Paris (1985)
27. Male, P.: L'adolescent et l'usage des drogues-point de vue psychologique. L'Information Psychiatrique **50**, 689 (1974)
28. Marcos, I.: Le Sens Urbain: La Morphogenèse et la Sémiogenèse de Lisbonne. Université Aarhus, Thèse de Doctorat, Aarhus (1996)
29. Marcos, I.: Morphogenesis of urban "glocalization". Con Moto ma non Presto–For Per Aage Brandt. In: Holsting, A., Mey J. (eds.) In RASK International Journal for Language and Communication, pp. 293–314. Odense, University of Southern Denmark, n°40. ISSN 0909-8976 (2014)
30. Marcos, I., Bayle, C.: Controverses: théorie vs praxis? dans Géopoint 2014: Controverses et géographies. Université d'Avignon, Avignon, En Ligne (http://www.groupedupont.org/ColloqueGeopoint/Geopoint14/Documents/GP14_PropositionsDebat_Web/GP14-A3-2-Bayle-Marcos.pdf) (2014)
31. Marcos, I., Morier, C.: La théorie sémiophysique de René Thom permet-elle de comprendre autrement le concept de frontière? In : WorkingPapers. MIS–Université du Luxembourg, [En Ligne] (http://mis.uni.lu/fr/wp-content/uploads/sites/7/2016/05/WP_Mise-en-frontière_Marcos_Morier_MIS-Working-Paper.pdf) (2016)
32. Marcos, I., Farate, C.: Corps, Territoire et langage: l'empreinte des origines dans l'actualité. (non-published) (2021)
33. Ritchot, G. et al.: Rapport d'études sur le patrimoine immobilier de Montréal. Centre de Recherche en Innovation Urbaine, Université de Montréal, ministère des Affaires culturelles du Québec (1977)
34. Soler, C.: Du parlêtre. L'en Je lacanien **2**(11), 23–33 (2008)
35. Thom, R.: Apologie du logos. Hachette, Paris (1990)

36. Zeeman, C.: Catastrophe Theory: Selected Papers 1972–1977. Addison-Wesley, Cambridge (1977)

Part III
René Thom Between Psychology and Biology

Chapter 8
The Notion of Archetype in Jung and in the Mathematical Theory of Rene Thom

Jacques Viret and Clement Morier

An archetype is etymologically a primitive model.[1] It is indeed generally considered as a theoretical reference model comprising the essential attributes, features or properties that characterize an object. It can be the model of a curriculum vitae for example, or the stylized model of the leaf of a tree drawn by a child. In biology, the archetypes represent, almost in the same spirit, the primitive or ancestral organic types common to several species and which are preserved through their differences. The wing of a bat, for example, has the same archetype as the upper limb of a human. In Plato, the word archetype applies to ideas, considered as immaterial and eternal models of things.

[1] Article already published in French in *Jung et les sciences*, Texts of the international conference « Jung et les sciences», Université libre de Bruxelles, March 21–22, 2009, E.M.E. editions, Fernelmont, 2009, p. 97–117, under the title Jacques Viret, « La notion d'archétype chez Jung et dans la théorie mathématique de René Thom», ©Academia-L'Harmattan. This English version was translated by Jacques Demongeot. Let him be warmly thanked here. *May this article celebrate the memory of our friend and mentor.*

Jacques Viret (1943–2018)

J. Viret
Pontcharra, France

C. Morier (✉)
Jean Moulin Lyon 3 University, Lyon, France
e-mail: clementmorier@yahoo.fr

© The Author(s), under exclusive license to Springer Nature Switzerland AG 2024

I. Marcos and C. Morier (eds.), *The Relevance of René Thom*, Lecture Notes in Morphogenesis, https://doi.org/10.1007/978-3-031-54982-3_8

8.1 The Notion of Archetype in Jung

If we refer to the work *The Vocabulary of Jung* elaborated by several specialists of the work of C.G. Jung (Agnel et al. [1]), the archetype is an empty structure acting as a virtual matrix, which is at the origin of a certain type of images, ideas or behaviors.

These are in fact patterns that are not invented but that can be encountered in myths, tales, fantasies and psychotic delusions, and which have an *energy intensity* that can lead to phenomena of fascination, even possession. According to Jung, archetypes are inseparable from the collective unconscious of which they precisely constitute the structure. In the course of his work, Jung progressively evolved his concept of archetype from a more or less fixed model, which he first calls *imago* or primordial image depending on the experience of the subject (images of the mother, the great mother, the wise old man or the divine child) to the idea of a model of process which conditions the behavior of the subject, a process closely linked to his energetic notion of *libido*. In this sense, as Aimé Agnel points out, archetypes can be defined as forms that can be grasped, through inner observation, of the a priori order in the psychic domain. Finally, the fact that archetypes are, according to Jung, forms of representation that constitute the structure of the psyche, has often been interpreted as possibly being hereditary. This is false, Jung often repeats, it is not the forms in themselves that are hereditary, but only the ability to represent them.

Thom, for his part, speaks little of the archetype, which he willingly considers in the primary sense of the term. However, he speaks of pre-program when he refers to the first tools made by men [9]. In fact, he says, referring to Bergson, the intuitive imagination which enabled homo *faber* to construct, long before any established science, tools endowed with remarkable efficiency, might this intuition not find its origin in a biological unconscious bequeathed by the phylogenetic evolution of species? The function creates the organ, not in the banal sense of an organic structure being created or developing by the use that is made of it, but in the sense that all the epigenetic and behavioral regulation of a species rests on a formal structure of geometric (topological) nature which is realized in the metabolic space of the organism. A function then appears as a partial regulatory device relating to the homeostasis of an essential physiological parameter, and all the functions, in the space of metabolic activities, thus contribute to the channeling of a dynamic, forming a global attractor, which constitutes a figure of regulation of the considered species. Finally, he added that as the global description of such a figure is out of reach for the moment, we can however hope to give local images of it in spaces of lower dimensions.

To each his language, his discipline, his way of conceiving and expressing himself, of course, but in the end, aren't these two men saying the same thing?: empty structures, forms of representation inherited over the course of human evolution for one and a formal structure of geometric shape, bequeathed by evolution in the phylogenetic sense of the term for the other; universal process which conditions behavior individuals for one, figure of physiological regulation of a given species for the other; motifs possessing an energy intensity linked to the Jungian notion of libido for one, channeling of a dynamic for the other.

8.2 Libido and Dynamical System

For Jung, the energetic intensity that he attributes to archetypal motifs is, as we have said, closely linked to his notion of libido. However, this notion, developed in his book "Psychic Energy" [3], remains relatively vague insofar as he defines it sometimes as an energetic flow and sometimes as an energetic or vital potential. From this point of view, Thom, more precise, speaks of a dynamical system [8]. What does that mean? A simple example of a dynamical system could be a *landscape* (which is actually a surface developed in a three-dimensional space), on which many drops of water flow from, for example, regular rain (Fig. 8.1). This landscape constitutes, in the mathematical language of Thom, a variety (in three dimensions) and similarly, all the drops of water which run off this relief can be considered as a field of vectors. Still mathematically speaking, a dynamical system is indeed defined as "the action of a vector field on a manifold". Depending on the relief, these dripping water drops gather in rills, then streams, torrents and rivers, the latter temporarily accumulating in lakes where the water seems to be at rest. The ups and downs of the relief make it possible to locally define a potential energy (here gravity) while the slopes taken by the water provide access, just as locally, to flow velocities or flow rates. When Jung speaks of the *libido* in terms of energy potential, we can think that he is referring to a potential energy (due to what he calls pairs of opposites, like here the top and the bottom of the relief), and when he talks about it in terms of energy flow, he seems to be referring to flow velocity.

To continue this example (or this metaphor), the speed of the water flow is canceled at the level of the lakes. In the language of Thom, these lakes where the speed cancels out are attraction areas (we will simplify by speaking only of *attractors*), the valleys

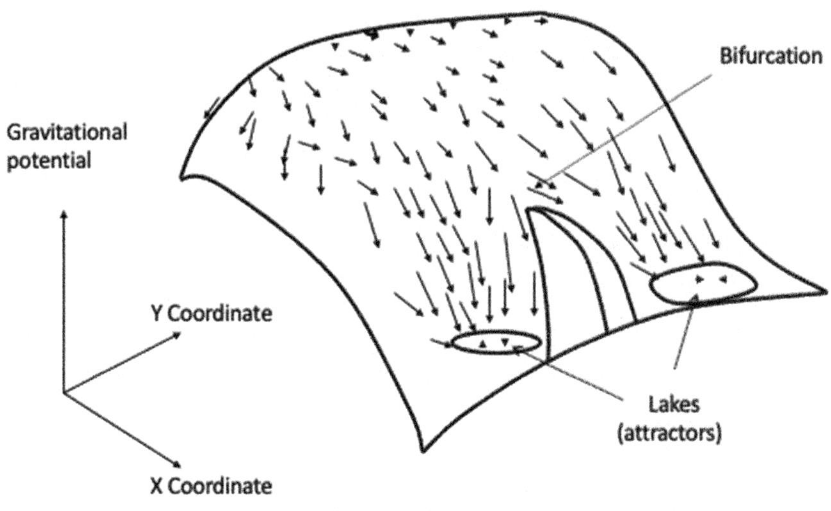

Fig. 8.1 Morphogenetic landscape with indication of two attractors and one bifurcation singularity

leading to these lakes being *attraction basins*. Attractors are domains of stability. Indeed, imagine that we observe this landscape for a sufficient time so that the effects of erosion can manifest themselves, then the high points (mountains or hills) will gradually disappear, but the low points (the lakes) will not will hardly change. Only the lakes located at altitude will be able to change place (therefore of domain of stability) because of the modifications of the beds of the torrents due to the movements of land following this erosion. In this case, Thom speaks of *bifurcation* of attraction basins, and of *catastrophe* when there is a rapid change in the domain of stability. A catastrophe, in the mathematical (topological) sense of the term, therefore amounts to a sudden change of attractor in a dynamic system.

Ultimately, when Thom studies a dynamical system, he mainly identifies its domains of stability (its attractors), as well as the bifurcations of its basins of attraction. But what does Jung do? The latter is more sensitive to variations in flow velocities. He more willingly notes, in his patients, areas of instability, symptoms which "jump to the eye", which "hurt", in brief which are immediately observable, bearers of a speed of flow, or, as he says, of a strong energy flow. In this hydrodynamic metaphor of the landscape, we could simplify by saying that if Thom rather observes the lakes and the relationships that they can have between them, Jung observes, him, rather the torrents and the waterfalls.

8.3 The Catastrophes of Thom

The sets of catastrophes (or sets of bifurcations) of Thom are like limited parts of a landscape, where a restricted number of attractors come into action. It describes seven elementary catastrophes that are enough to model what is happening in our macroscopic world. These catastrophes can be formalized by polynomial mathematical functions, called "deployments", an aspect on which we will not insist.[2] We will only discuss the general shape of two of these sets of catastrophes, called respectively the "fold" and the "swallowtail" and which are illustrated pictorially in Fig. 8.2.

The *fold* is not strictly speaking a catastrophe set since it includes only one attractor. The latter corresponds to the potential depression of curve (Fig. 8.3A), where we see that if we give sufficient energy to the ball placed in this depression zone, it escapes from the curve. The top and bottom sheets of the fold in Fig. 8.3B represent the stability domains of the system, and the arrow indicates the trajectory taken during the disappearance of this domain. Let's imagine that this system is a boiler in stable operation, and that we push the heating pressure. There comes a time when the boiler explodes. This is a disaster in the usual sense of the term, of course, but once again, not in the mathematical sense since the boiler, destroyed, can no longer find another stable mode of operation (in this case, there is only one attractor).

[2] See Y.L. Kergosien, R. Thom. *Sur les points paraboliques des surfaces*. CRAS A (1980), 290, 705–710 (Translator's note).

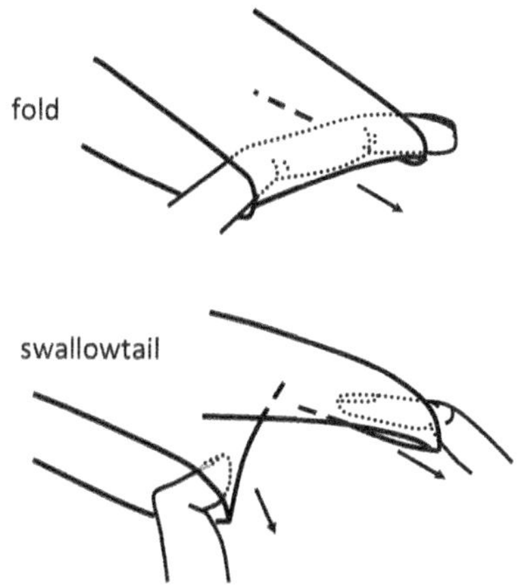

Fig. 8.2 The fold (top) and swallowtail (bottom). The representation of catastrophes uses fingers in a tissue (as suggested by J. Demongeot)

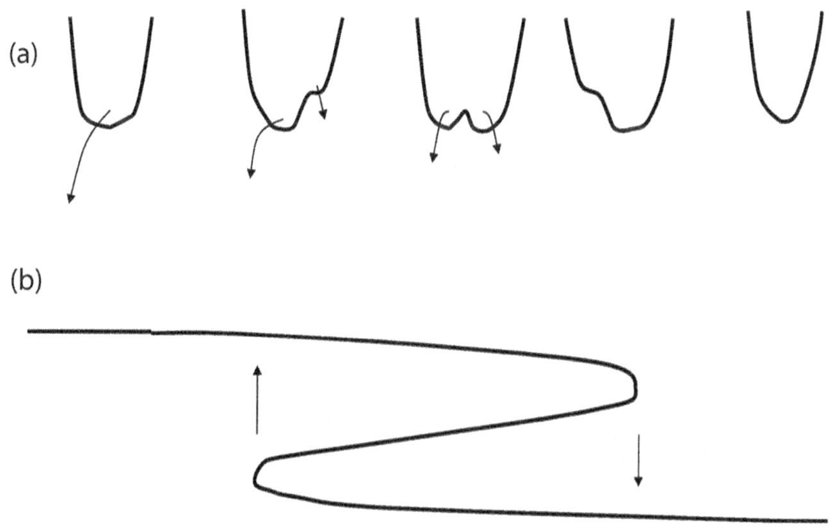

Fig. 8.3 The cut of the fold evolves under the effect of the dynamics of a parameter from a parabola with a minimum (A on the left) to a quartic with two minima (A in the middle) giving the catastrophe called *cusp*. On B, the top and bottom sheets of the cusp represent the stability domains of the system, and the arrow indicates the trajectory taken during their disappearance

8.4 Thom's Swallowtail and Jung's Psychic Structure

According to Jung [2], the psyche is structured into a collective unconscious, a personal unconscious and a conscious. The collective unconscious presents itself, according to him, as a vast sediment of psychic motives, not specific to an individual, gradually accumulated during the long human evolution, sediment on which rests a fine individual silt, made up of the personal unconscious and consciousness that

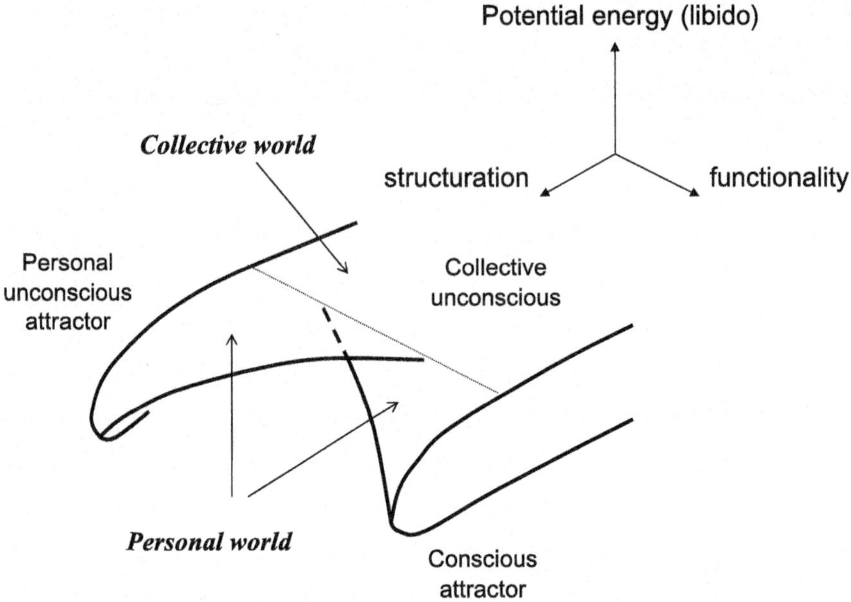

Fig. 8.4 Organization of the unconscious and conscious levels in the Thom's swallowtail representation

characterize a single individual. Thom's swallowtail, represented in Fig. 8.4, takes up this organization. One can observe there the collective, undifferentiated world, and the personal world where the two attractors "personal unconscious" and "conscious" gradually differentiate. This figure unfolds in a three-dimensional space, along three axes.

The vertical axis represents a potential that could be described as "vital", in the sense that an individual has, at birth, a certain capacity to develop his own personality. As everything starts with the collective unconscious, it is naturally the latter that has the greatest potential. It all depends, of course, on how that individual is going to use this potential, in other words, it all depends on the trajectory he is going to follow. On the horizontal plane, one of the axes represents the differentiation (structuring) of personal conscious and unconscious attractors, while the other describes the possible relationships (functionality) between these two attractors. In the personal world, there is an overlap zone of differentiated attractors.

As shown by the vertical sections of the swallowtail made in Fig. 8.5, this zone represents a possibility of "dialogue" (figured by a trajectory) between the personal unconscious and the conscious (section B). When the two attractors, admittedly differentiated, are further dissociated (section A), there is no longer any possibility of dialogue, no longer any possible functional trajectory. The conscious or (and) the personal unconscious are in predominant positions. In this case, Jung says of the personal unconscious that *it dominates the subject* and of the conscious that it is *in*

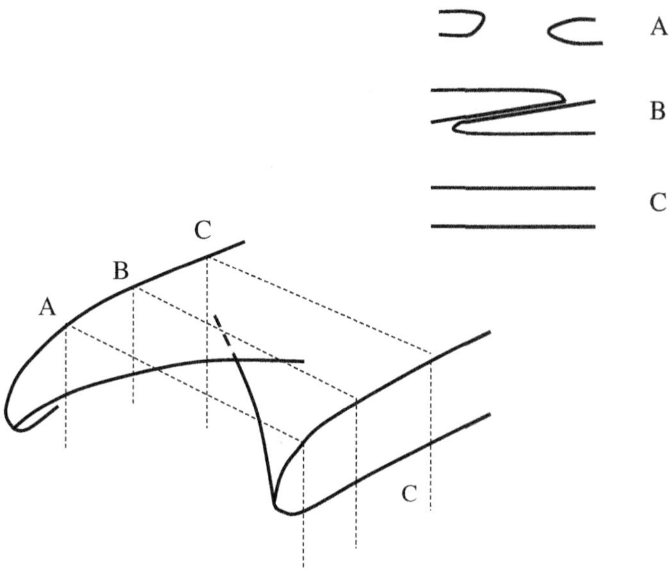

Fig. 8.5 Three successive vertical sections of the Thom's swallowtail surface

a too unilateral position. Naturally, there cannot exist any possibility of dialogue in the C cut since in the collective world, the attractors are not yet differentiated.

8.5 Psychic Function in the Biological Sense of the Term

A biological function can be described by a cyclic trajectory with four phases, deployed on the swallowtail [9, 10]. If we accept the term "psychic function", we can schematically describe the integration of an unconscious content into the conscious through such a trajectory (Fig. 8.6). We will take as an example an individual who, for various reasons, is in a state of what Jung calls a *period of regression*, that is to say a state of blockage of the natural energy flow of the libido. This blockage leads to a kind of stasis, caused by an accumulation of unconscious products not integrated into the conscious, and to a personality increasingly under the influence of the unconscious which limits the capacity of the individual to adapt psychically to the present difficulties of his environment. This overload of libido then forces the unconscious to "react" by activating deeper contents, often considered as infantile, but in fact archaic [6]. This could be the case, for example, of a subject of the predominant thought function type, solicited by a requirement of life of the order of feeling and well decided to solve the problems posed by this solicitation only by a rational reflection.

The function cycle begins in Fig. 8.6, as in any function in the biological sense of the term, with a structural phase (a) which culminates in the perception of an

a = structuration, excitation, rigidification
b = biological act } Action

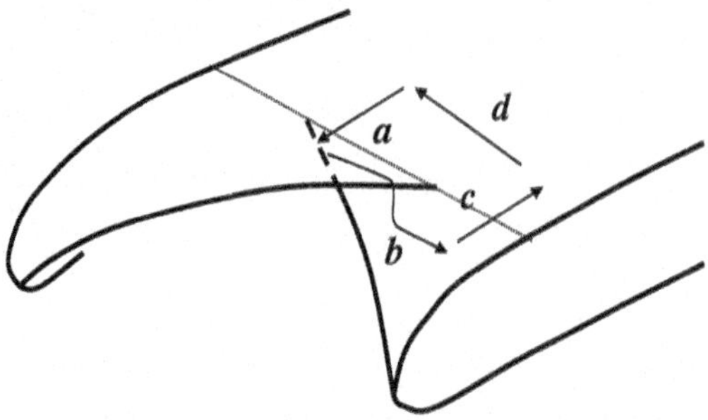

c = relaxation, inhibition, thinning
d = assessment of the act, investment of the next } Refractory state

Fig. 8.6 Biological function described by a cyclic trajectory with four phases, deployed on the swallowtail

archetypal psychic image. Here, the structuring consists for the unconscious, not in really organizing itself but rather in looking for a breach in this blockage, and this with the aim of opening an energy flow path for this excess of libido. This breach can only exist at the level of the overlap zone of the attractors, the only possible way out of the bifurcation towards the conscious. When, in Fig. 8.6, the trajectory arrives at the vertical of the lower layer (of the conscious attractor), there is, in the language of Thom, a catastrophe of perception. The suddenness of the perception of this irruption of the unconscious justifies, moreover, the appellation of catastrophe. It is during the second phase (b), the functional one, that this unconscious content can be recognized by the conscious of the subject (generally with the help of the therapist). This is, according to Thom, a catastrophe of capture, of action. The third phase (c), again structural, is a phase of relaxation, marked by the emotion that the conscious recognition of this unconscious content gives to the subject. It is usually a strong moment, often marked by fascination, but sometimes by anger, depending on the nature of the unconscious content. Finally, the fourth phase (d) corresponds to the complete integration, in other words the comprehension, of the previously conscious content at the same time as a return to the initial potential level of psychic energy. This last phase is slow (it is a period refractory to action), and it is more "energy dependent". It is indeed not always very easy to integrate, that is to say to "live with", an aware content even if it is perfectly identified. In return for this energy expenditure which allows the recovery of the initial potential, the subsequent integration of new content will be facilitated by the implementation, then the repetition, of these functional cycles. Ultimately, the last phase of a cycle should be viewed primarily as a functional investment phase. Communication between the

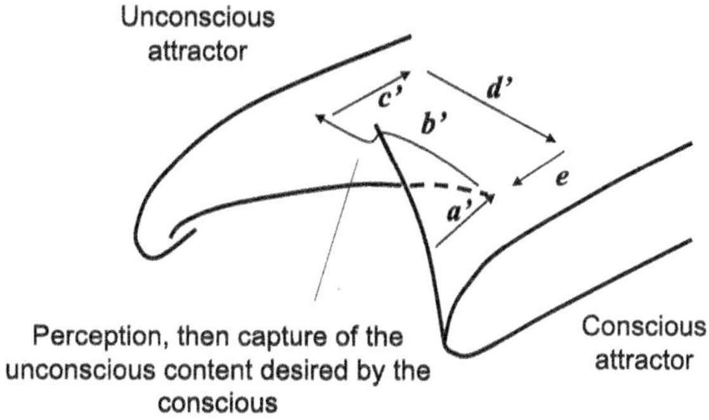

Fig. 8.7 Progression of psychic functions from a conscious state (a') to an unconscious one through the capture of unconscious content (b') until the return to the conscious state (e) through a zone of attractors overlap (c'–d')

unconscious and the conscious can then become more harmonious, and a process of progression (in the Jungian sense of the term) can then begin or resume.

In the progression, the psychic function begins at the level of the conscious attractor which is this time in a potential position superior to that of the personal unconscious (Fig. 8.7). This conscious attractor must, during the first phase (a'), destructure itself, dedifferentiate itself retrogradely to the structural axis, abandon in some way its natural defense against the irrational and this until it comes to place itself in a condition to perceive what comes from the unconscious, that is to say up to the zone of overlap of the attractors. The second phase (b') corresponds to the capture of perceived unconscious phenomena. This is really a capture in the predation sense of the term. It is indeed the conscious that solicits this capture, knowing all the same that the unconscious contents are not identified a priori, just as the nature of a prey is always a surprise for a predator. It's up to him to deal with it. Phase (c') corresponds, in the case of progression, to a reduction in the tension of the unconscious, tension which had accumulated during the period of previous regression, and during phase (d'), the unconscious collective should experience rearrangements of archetypes of which, of course, we do not know the nature. Finally, phase (e), which ends the cycle, could correspond to the preparation of an image modulating perhaps the previous content. In fact, the process described above is essentially what happens during what Jung calls the *active imagination*.

8.6 The Ego and the Process of Individualization

The overlap zone of the two swallowtail attractors therefore appears as a dialogue zone between the conscious and the personal unconscious. This is where the subject's true personality, ultimately his *ego*, is formed. Figure 8.8 shows a cross-section of this region, showing the two overlapping attractors. The unconscious and the conscious are located on either side of an oblique line which acts as a border between these two entities and can be considered as their "threshold". In fact, this line is formed by the connection of the "edges" of the two leaves representing the conscious for the superior

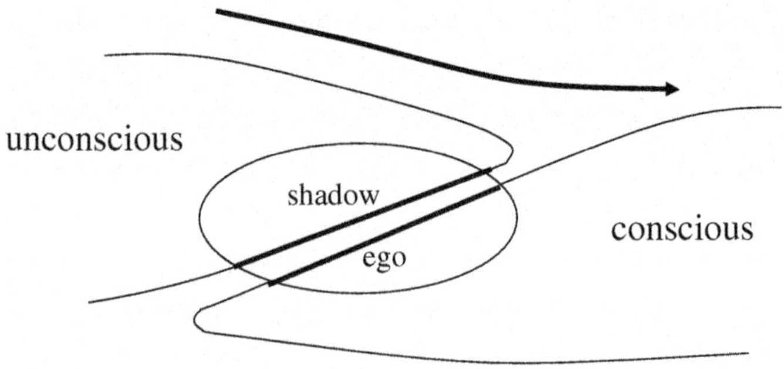

Fig. 8.8 Cross-section of the region of attractors overlap, unconscious and conscious being located on either side of an oblique line which acts as a border between these two entities and can be considered as their "threshold"

and the unconscious for the inferior. More exactly, the *self* would be located on the upper edge of the lower sheet, therefore of the conscious, while the homologous part located on the lower edge of the upper sheet, therefore of the unconscious, would correspond to "the shadow of the Ego".

From a topological point of view, the self and the shadow of the self can be considered as projections, on "thickened points", of the totality of the lower and upper ranges, themselves resulting from the accumulation, so as not to say of the integration over time, of conscious and unconscious phenomena. In reality, we know that the shadow corresponds to the aspects of our personality that the conscious rejects or at least does not accept. This archetype therefore only appears with the differentiation of consciousness, that is to say with the advent of the self, an entity of which it constitutes the antagonistic polarity. Most of the time experienced as negative, it can however, when activated by a phenomenon of self-compensation, play a positive role by allowing the individual to glimpse his true "potential". The expression "shadow of the Ego" was taken up and developed by Solié [7].

8.7 Back to the Swallowtail

In reality, the morphology of Fig. 8.6 unfolds in a four-dimensional space and not three as we have represented it so far. As it is impossible to carry more than three dimensions on a diagram, and if one wishes to express a new dimension, it is necessary, on the other hand, to delete an old one. In mathematical language, we say that we hide a variable to allow the expression of a new variable hitherto hidden. If we therefore hide the "psychic potential" variable, we obtain a plane figure which is in fact the projection (we could also say the signature) of the swallowtail on the lower plane (Fig. 8.9), provided only with the "structure" and "function" variables.

Then, by showing a new variable (hitherto hidden) of the swallowtail, we obtain the morphology of Fig. 8.10, which is a new representation of this same swallowtail.

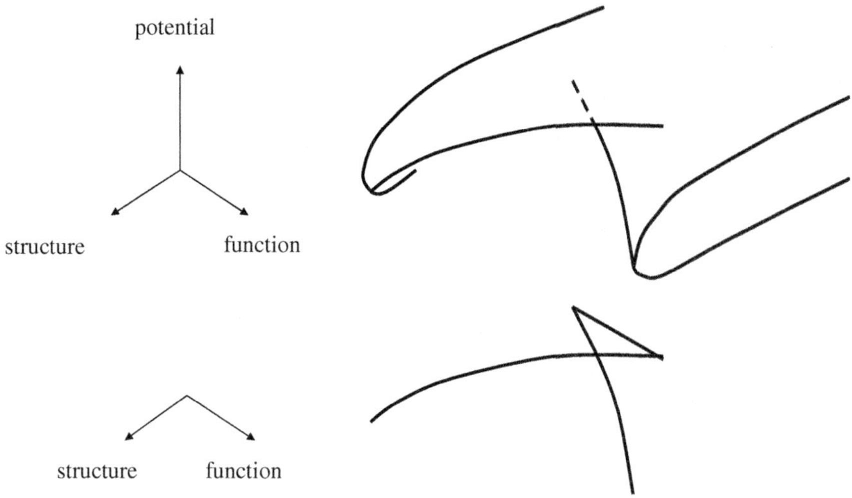

Fig. 8.9 Top: three-dimensional morphology of Fig. 8.6. Bottom: Projection (or signature) of the swallowtail on the lower plane, provided only with the "structure" and "function" variables

This new surface describes, in fact, the place of the points for which the minima and the maxima of the psychic potential are reached, according to this new variable which one could qualify as *evolutionary parameter*. This surface has in fact three distinct regions that we will name A, B and C. Region C is the one that we already know as being the projection of Fig. 8.9, with in particular the bifid zone of overlap of the two attractors. Before B, one observes an angular point which constitutes in fact the germ of this future bifid zone, and in A, one does not distinguish any particular point, but only a harmonious or regular curve. To sum up, we can say that the progression from A to B is accompanied by the emergence of the two attractors, and that from B to C marks their development. There is therefore an evolutionary significance to this new variable.

Figure 8.10 allows us to summarize this evolution, with, juxtaposed, the morphologies deployed in the old three-dimensional space. We find, in C, the swallowtail, while in A, it is the fold. From a mathematical point of view, this figure makes it possible to realize the topological continuity of the fold to the swallowtail by the removal, between A and B, of the functional variable, hidden for the fold, present for the tail of swallowtail (this means that the swallowtail contains the fold). From the psychological point of view, it is a question, between regions B and C, of what one could qualify as the development (or progression) of the ego, that is to say the unceasing deepening renewed dialogue between the personal unconscious and the conscious. The conscious gradually nourishes itself by assimilation, then integration, of the contents of the unconscious (we have spoken of the shadow, but it is the same for the persona, the animus and the anima), and this leads to the development of the consciousness and the personality. This is the process of individualization.

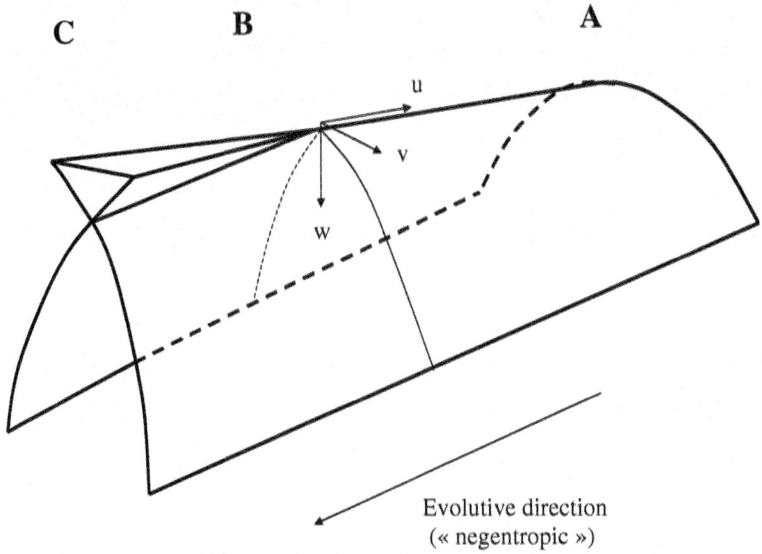

Fig. 8.10 Due to the increase of a hidden variable, the evolutionary parameter, the progression from A to B is accompanied by the emergence of two attractors, and that from B to C marked by their development

8.8 Evolution Variables

To what can correspond, in Jungian psychology, the variables evolving according to the three dimensions of the swallowtail of Fig. 8.10, called (u), (v) and (w) in the nomenclature of the theory of catastrophes of Thom?

The variable (w), as we have already seen, is structural in nature. It corresponds in fact to the structuring of the "personal unconscious" and "conscious" attractors, in other words to their differentiation. When it remains moderate, it describes a zone of overlap of these attractors, with the possibility of a passage from one to the other. It is, of course, the dialogue between the personal unconscious and the conscious, in other words the ego. When it is high, it describes the dissociation of attractors which then evolve on their own account. This is where Jung speaks either of a *conscious in a too one-sided position*, or of a *personal unconscious that dominates the subject*. However, a closer observation of Fig. 8.10 shows that the increase in the values of (w) is accompanied by a spatial rapprochement of the attractors, even dissociated ones, as well as an increase by "stretching" of the surface of the dialog box (remember that topology is often defined as the geometry of elastic surfaces). The increase in the dialogue surface can be interpreted as a progression of the ego, while the bringing together of the attractors makes it possible to envisage a facilitation of the episodes of *regression* described by Jung, and which we have seen that they also participate, whatever indirectly, to the progression of the self. Ultimately, the variable (w) seems closely linked to the evolution of the ego.

In Fig. 8.10, the variable (v) is functional in nature. In one direction, it describes the fact that action, i.e., the emergence of unconscious content at the conscious level, is a process of catastrophe (in the *Thomian* sense of the term), therefore rapid and spontaneous, often experienced by the conscious as if not numinous at least manifestly new. We can say that during this process, the conscious is then turned towards this new object "external" to itself and, in this sense, we can qualify this phase as extraversion. The other direction, which we have called refractory, describes the very slow process of integrating this content into the conscious self. It generally results from an "inner" personal investment work, and therefore seems to correspond to a phase of introversion.

Variable (u) was initially qualified as an evolution variable. However, its origin is at the angular point of the region B, and its values increase in the opposite direction to that of the unfolding of the swallowtail morphology towards its bifid region. It physically acts as a return spring that retains the system in its initial state of equilibrium. Its psychological translation seems to be to slow down the differentiation of the personal world and therefore of the ego (let us remember in fact that it is only when it is canceled at the angular point that the personal world can be born, that is to say the awareness). As Jung reminds us during his seminars on the theme of children's dreams [6], "It is in the unconscious that our evolution begins. If we don't understand this, we forget that we came from the animal kingdom. It is indeed for these children to tear their self from the original world". By opposing the complexity of psychic evolution (what Solié [7] calls negentropy), this variable (u) can be considered as an entropy variable, the cancellation of which marks the border where animal life becomes human. From this point of view, man may have only become man because he was able to overcome the natural resistance to remaining strictly animal, because he knew how to make the values of the variable (u) negative. If we leave this phylogenetic point of view to come to the individual, this stage is probably a necessary passage from childhood.

8.9 The Limits of the Swallowtail… And of the Ego

The swallowtail is therefore a geometric model that illustrates the process of individualization, in other words the development of the self. Not only does this ego belong to the personal world, but it is also the center of consciousness, constantly reinforced (according to the variable w) by the projection of the points which constitute the common border between the two attractors "personal unconscious" and "conscious", at the level of their overlap zone. However, if, because of the progression of this recovery, the ego does indeed carry within it this increasingly strong and balanced image of the center of consciousness, it also carries the germ of a possible imbalance, this time due to the excessive differentiation, even the possible dissociation of these two attractors during this progression. As can be seen in pathology, this can then lead to dissociation from the ego.

Moreover, if the variable u, of an entropic nature, is opposed to the consciousness evolution, one can wonder if there is not another variable, of a negentropic nature this time, which, as said P. Solié, could account for the growing complexity of psychic evolution. Vital energy, in the broadest sense possible, that is to say the need to survive, is certainly necessary for this survival, but it does not on its own explain the channeling it must undergo in order to succeed effectively to this complex evolution. But the swallowtail does not allow us to access such a variable. However, Jung answers this question by saying that beyond the process of individualization, there is that of individuation. Beyond the ego, there is the central archetype of the self.

8.10 Topological Approach to the Self

In the context of this presentation, we will only address this question which would in reality require a long development. For Jung, the self is the archetypal center of the total personality, conscious and unconscious. It represents the union of all opposites, and in the first place, that of the ego and the shadow of the ego (antagonistic shadow). The shadow of the ego, by definition one might say, is most often projected onto others. The withdrawal of this projection, along with the subject's recognition that this shadow is actually part of him (actually the self), is accompanied by a defeat of the ego, a distortion of the ego. It is the same for the projections (always unconscious) of the anima of the man on the woman or of the animus of the woman on the man. It is only, as Elie Humbert says, with the withdrawal of these unconscious projections onto others, and the bringing together of opposites that this withdrawal implies, that "*the ego can recognize the self as the true axis of growth of the psyche*". The experience of self-realization is always painful. The union of opposites, which he pushes to achieve, is therefore not spontaneous, but most often provoked by serious internal disorders which, leading to an intrusion of the collective unconscious, first affect the ego. On the other hand, the realization of the self allows access to a totality, to what Jung calls the whole man.

There is, in catastrophe theory, a set of bifurcations which could be said to follow the swallowtail set, when new constraints arise. This is the umbilicus. Figure 8.11 represents a section of the region C of the swallowtail, assailed by a shock wave coming to constrain, in D, the overlap zone of the attractors. An opening is then created (at E then F), which leads to the formation of two lips (at F). The disappearance of the overlap zone of the attractors leads secondarily to the disappearance of the dissociated part of the latter (in G).

If we come back to psychology, we can consider that the shock wave consists of the intrusion of the collective unconscious into the personal world, which it jostles and opens up at the level of the dialogue zone between the conscious and the personal unconscious, thereby resulting in a distortion of the ego (Fig. 8.12).

The two lips formed during this opening of the self are in fact the active traces that remain of the old attractors, on the one hand the conscious and on the other the personal unconscious. They delimit in fact a kind of cavity, a veritable crossroads

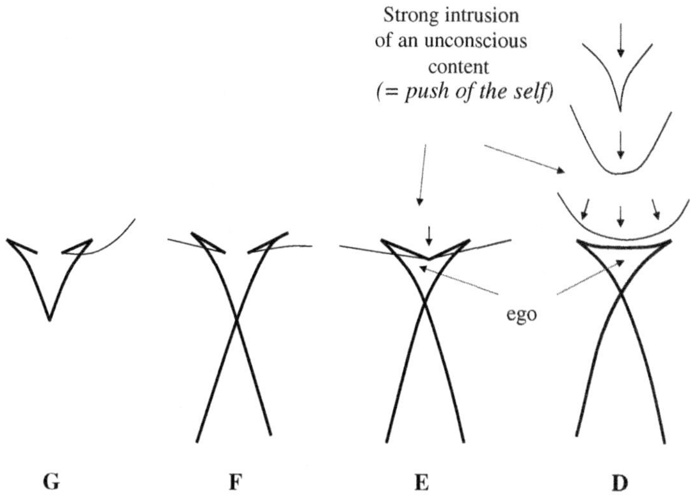

Fig. 8.11 Sections of the region C of the swallowtail assailed to constrain, in D, the overlap zone of the attractors. If we retrace this sequence by juxtaposing all the sections (from D to G), we obtain on Fig. 8.12 the bifurcation set of the umbilicus

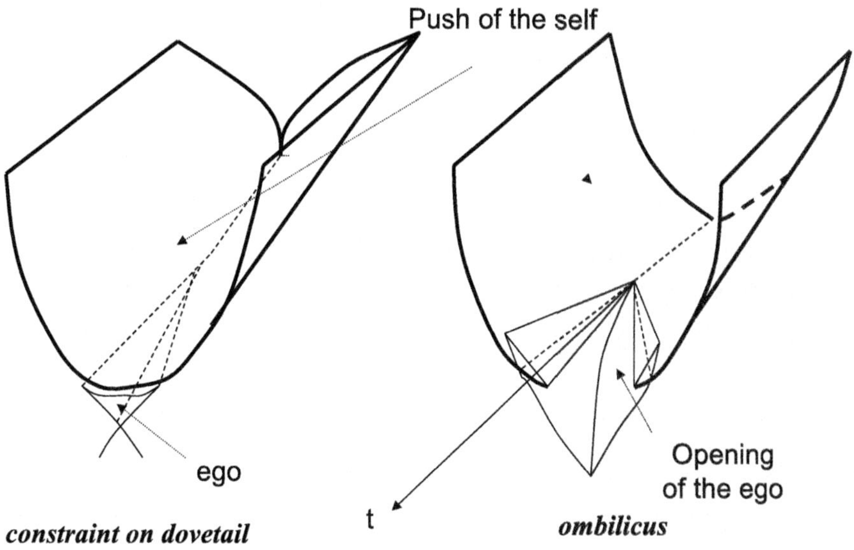

Fig. 8.12 Bifurcation set of the umbilicus appearing in the sequence juxtaposing all the sections of Fig. 8.11 (from D to G)

swirling under the dependence of the energy conveyed by the self and seat of an activity where conscious and unconscious are forced to work in synergy (union of opposites), the self, even weakened, remaining the only conscious will in this difficult dialogue with the collective unconscious. This morphological space could be what Jung calls *the ideal place for the symbols of a synthetic and unified view of the psyche to arise* [5].

Finally, the umbilicus being a morphology of dimension greater than the swallowtail, its deployment reveals a new variable, named t. In catastrophe theory, its origin is located at the level of the opening of the zone of overlap, that is to say at the level of the cut E of Fig. 8.11. This corresponds, in the psychic transposition that we have proposed, to the thrust and intrusion of the collective unconscious into the personal world, that is to say, according to Jung, to the beginning of self-realization. We can therefore think that the variable (t) corresponds to this realization, that is to say to the evolution of individuation.

8.11 To Conclude

- *On the notion of archetype*: Despite what is announced in the introduction on the idea that Jung has of the archetype as being able to be a model of process which conditions the behavior of the subject, the latter seems to remain very attached to his concept. empty structure that collects unconscious contents, then presents them, thanks to this structure, to the personal world in the form of images in which conscious and unconscious contents can be linked, this constituting "the best possible expression for some something unknowable, or which is not yet knowable", in other words a symbol [4]. The vision given to us by the morphologies of Thom's catastrophe theory concerning the archetype would rather be that of a path to be taken, or more exactly of a progression, which is more related to the notion of process, therefore of evolution. However, the two authors agree on the importance of the energy brought into play of factualization), for Thom, it is the dynamical system which makes it possible to transform potential energy into factual energy (only the factual aspect is taken into consideration).
- *On the morphologies*: There is, it seems, perfect concordance between on the one hand the process of individualization (development of the ego) of Jung and the set of bifurcation of the swallowtail of Thom, and on the other hand between the process of individuation (realization of the self) of Jung and the set of bifurcation of the umbilicus of Thom. It is in this sense that Thom speaks of "the epigenetic and behavioral regulation of a species which is based on a formal structure of a geometric (topological) character". However, let us point out a small problem concerning the evolution variables. The variable (t), of which we have said that it took its origin at the rupture of the zone of overlap of the attractors and which seems to describe the process of individuation, should, despite everything, be considered as the true negentropic variable (in the sense that gives him Pierre Solié), thus opposing the variable (u). Now in this case, there is a hiatus between

its point of origin and that of u, since this last variable cancels out at the angular point of the swallowtail. We find the same problem in Jung, who describes the process of self-realization after that of the development of the ego. This problem could be solved by considering, in catastrophe theory, that the umbilicus contains (in "hidden" form) the swallowtail (just as the swallowtail contains the fold) and that in Jungian psychology, the process of individuation begins at the same time as that of individualization, but that it evolves at the beginning in a hidden form. Wouldn't what the Jungian school calls "*the manifestations of the self*" in reality, not only manifestations, but an early attempt, if not at realization, at least at early liberation of the latter?

References

1. Agnel, A., et al.: Le vocabulaire de Jung. Ellipses, Paris (2005)
2. Jung, C.J.: Dialectique du moi et de l'inconscient. Gallimard, Paris (1964)
3. Jung, C.J.: L'énergétique psychique. Georg et Cie, Genève (1973)
4. Jung, C.J.: Réponse à Job. Buchet/Chastel, Paris (1977)
5. Jung, C.J.: L'âme et le Soi. Albin Michel, Paris (1990)
6. Jung, C.J.: Les rêves d'enfants. Albin Michel, Paris (2002)
7. Solié, P.: Le sacrifice. Albin Michel, Paris (1988)
8. Thom, R.: Modèles mathématiques de la morphogenèse. Christian Bourgeois, Paris (1980)
9. Thom, R.: Apologie du logos. Hachette, Paris (1990)
10. Viret, J.: Generalized biological functions. Acta Biotheor. **53**, 393–409 (2005)

Chapter 9
Psychoanalysis and Catastrophe Theory

Benoît Virole

Soon twenty years will have passed since the death of René Thom (1923–2002), not much in terms of the history of ideas, but enough to be able to evaluate today, without any fashionable issues, what is the scope of catastrophe theory in psychoanalysis? We shall understand this word "scope" in a double sense: scope by extension—does the theory of catastrophes present a possible extension in psychoanalysis in its clinical dimension as well as in its theoretical dimension—but also scope in the sense of support. To what extent could the theory of catastrophes support a new impulse of psychoanalysis?

9.1 The Theory of Catastrophes

In the theory of catastrophes, any domain that we wish to study is assimilated to a dynamic system whose exact nature is unknown but that we can observe in its phenomenological manifestations (its apparent states). We consider this system as likely to take a certain number of stable states generated by an unknown internal dynamics. These states are described in a set of descriptive variables of the system (state variables in the space of phases taken by the system). By controlling the values of other variables, called "control parameters" or "control factors", which vary in another space (external space), the apparent states of this system are modified. Variations on the control parameters trigger changes in the system's states. This minimal formulation is applicable to all kinds of systems, whether they are physical, biological or psychological, provided that they are subject to a gradient dynamic tending towards energy minima and that the variables are correctly identified.

B. Virole (✉)
University of Paris, Paris, France
e-mail: benoit.virole@wanadoo.fr

© The Author(s), under exclusive license to Springer Nature Switzerland AG 2024

I. Marcos and C. Morier (eds.), *The Relevance of René Thom*, Lecture Notes in Morphogenesis, https://doi.org/10.1007/978-3-031-54982-3_9

Although the pleasure principle reigning in the unconscious can be assimilated to an energetic gradient oriented towards its minimum, it is difficult, in psychoanalysis, to define the system on which the catastrophic model would apply. Is the system the whole psyche, instances included? Is it the unconscious? Is it the ego, organized instance, of which the superego would be the space of control, and the defenses its regulations? Is the system circumscribed to the development of the libido, with its phases, its integrations and regressions, its fixations, its sublimations…? And if we arbitrarily choose to pose the system as one or the other of these alternatives, what is the theoretical gain of such a construction? Don't we risk obtaining in output what we have inserted in input?

9.2 Difficulty of Application in Psychoanalysis

The direct application of the "catastrophic" method to the psychic functioning risks moreover to objectify the psyche of the patient. In psychoanalysis, we are not in a situation of observation of an external system (for example, the psychic complexity of the patient with its internal pulsional dynamics and the superego as a control space) but in a situation of intersubjectivity. The system considered should therefore be the couple associating the psychoanalyst (his psychic flow of thoughts and affects, his mental states, his counter-transference) and his patient (his psychic flow of thoughts and affects, his mental states, his transference). But then, it becomes difficult to define the control space, the number of factors, the number of state dimensions, and the topological type of the bifurcation set. A search for catastrophic modeling would thus be an intellectual construction not only without concrete interest, but moreover heterogeneous to the technique of psychoanalysis. The latter imposes the release of preconceptions and must leave room for the unknown. In spite of these reservations, certain contributions of the theory of catastrophes seem to us to be interesting elements for psychoanalysis. We will distinguish three of them. The first is centered on the technique of conducting a treatment, the second on the understanding of the relationship between the unconscious and writing, and the third on the structural dynamics of identifications.

9.3 Attention to Critical Moments in the Treatment

In the course of the session, we encounter, through the patient's discourse, apparent states of the self, in particular emotional states. These can be stable but they also encounter instabilities that denote the existence of an internal dynamic. This dynamics is not intelligible by reactions to the environmental modifications of the patient's life, nor to the hazards of the concrete existence (they are only on the surface) but it is under the influence of internal factors of the psychological life. At the time, this dynamic is unknown both to the psychoanalyst and to the subject himself. It can only be apprehended afterwards. For example, in front of the sudden appearance of an anxious feeling whose intensity is out of any proportion with a real event, we will infer the existence of an internal dynamics of repression by the self of unconscious desires or reactivated traumatic motions. By translating this dynamic in terms of the theory of the catastrophes, the precipitation of a symptom or a substitutive formation results from the crossing of a stratum of a space of three-dimensional bifurcation, since it obeys the interactions between at least four dimensions, the two organized instances of the ego and the superego, the requirements of the id and those of the reality (social, family, cultural, economic). This space of complex bifurcation, not representable, allows to conceive—virtually—the appearance and the disappearance of psychological manifestations. This is only a metaphor, of course, but it is a technically operative metaphor. Thus, the catastrophic notion of bifurcations is associated with the notion of critical zones. As close as possible to the catastrophic singularity, there is a reactive sensitivity that can induce a bifurcation. One could then imagine the art of analytic interpretation as being that of detecting these critical zones, perhaps by intimate resonance between the mental attractors of the analyst and those of his patient. An intervention, given with tact, sometimes a simple comment, a punctuation, can exert a psychic action of great intensity and induce a progress in the treatment. The theory of catastrophes, used metaphorically, thus gives consistency to the therapeutic intuition: the proximity of a critical zone in the patient implies its detection by the analyst. It also involves the risk of an interpretation that will allow a bifurcation of meaning, a qualitative change in self-knowledge, in other words, an insight.

9.4 Repression and Form

René Thom teaches us to detect under any representation the existence of an agent scheme resulting from a dynamic of conflict. The pictogrammatic and ideogrammatic writings (archaic sinograms) thus reveal an obvious analogy with the schemes of elementary catastrophes. But the fundamental syntactic structures of verbal language are also direct expressions of elementary catastrophes. The nuclear sentence can be considered as a drama (Lucien Tesnière) between protagonists assimilated to attractors in competition. This catastrophic conception is illustrated magnificently by the semiotics of the gestural languages whose space of realization is quadri-dimensional (3 dimensions of the gestural space plus the time) thus avoiding the dimensional collapse inherent to the linearity of the oral languages where the pluridimensionality must be projected on the sequential chain of the words. The sign languages of the deaf reveal thus in their gestural iconicity the agent schemes of elementary catastrophes. The presence, under all representations of agent schemes, allows a reinterpretation of the use of metaphors (analogies).

If an object A has a morphodynamic core topologically identical to that of an object B, then A and B can be substituted in a metaphorical relation. Take the following metaphor, old age is the evening of life. The two signifiers "old age" and "evening" share the same morphodynamic pattern of termination that we represent by the following diagram —⊣ which can be read as, something exists — and then ends ⊣ . Now this morphodynamic scheme results from a fold catastrophe whose bifurcation space is a single point. Since these two signifiers possess a common scheme, any analogy is true in the sense that it reveals a similar dynamic link. This is what allowed Thom to say to Lacan, during a lunch where the psychoanalyst had invited him: what borders on truth is not error but insignificance, i.e. the absence of meaning. The formulation is provocative, but it is fundamental in the epistemology of catastrophe theory. By elevating analogies and metaphors to the rank of vectors of investigation of meaning, one generates a virtual space in which one can study the potential deployment of observed phenomena. Of course, the investigation of the patterns common to two objects does not mean that they are identical—we would then be in a delusional thinking—but these objects share a common dynamic genesis. The phenomena of convergence in biology, the identical forms of oil drops in water and the forms of jellyfish, constitute illustrations of these analogies, false as regards determinism by the substrate, true for catastrophic determinism indifferent to its nature.

The existence of an agent scheme existing under any mental representation, then, becomes a useful notion for understanding the symbolic equations realized by the unconscious. For example, the unconscious symbolically assimilates a penis to an object of oblong shape with an unfolding end, a vulva to an excavation, breasts to the contours of a hill, etc. The repression—defense of the ego—seeks the compromise between the unconscious requirements of the satisfaction of the sexual impulse coming from the id and those of the superego aiming at putting away the disturbing sexual representations. It is thus a search for compromise requiring the coexistence of contradictory elements. The application of the theory of catastrophes to the nature of the representation provides particularly interesting elements to conceive how this compromise can be carried out. The ego, repressing the disturbing representation, under the injunction of the superego, tolerates its dynamic "skeleton" because the sexual semes are displaced but the id finds satisfaction by the representation of the sexual impulse through morphodynamic features (penetration, excision).

It is thus common experience during an analysis to encounter in the dreams and fantasies of the patient, the evocation of strange forms, incomprehensible to him. These can be kinds of locks, pins, forks, and holes. The interpretation in terms of disguised sexual representations, formations of compromise, does not have to be questioned—the sexual is indeed consubstantial to the unconscious—nor their function of enigmatic signifiers as proposed by Jean Laplanche or of original pictograms for Piera Aulagnier. But it is insufficient if it does not integrate the dynamic processes that allowed their genesis. Catastrophe theory can thus shed new light on these phenomena that can be observed in clinics where the unconscious seems to have become a scribe. The Freudian metaphor of the unconscious as a "*Wunderblock*", a magic slate, allowing all psychic palimpsests, is singularly congruent with the knowledge of the agent graphs of catastrophes. The relevance of the concept of an unconscious structured like a writing (Jacques Derrida, Piera Aulagnier), full of indexical traces of complex dynamics, can thus only be reinforced by the contributions of the theory of catastrophes. On a technical level, this knowledge allows the analyst to infer a potential meaning on enigmatic objects invested by the patient and whose choice can be enlightened by the identification of its traces (points and forks, curves and splinters). On a speculative level, in the line of René Thom's models, one could conjecture that the whole of the development of the libido follows the unfolding of a multidimensional aggregative catastrophe of which the stages (oral, anal and phallic) would only be local sections and of which one would have knowledge by their indexical traces, and precisely by the inscriptions of their agent schemes in the analytic material. To conceive the development of the libido as a generalized catastrophe, with local sections corresponding to the partial drives, is also a powerful metaphor that evokes the grandiose construction of Sandor Ferenczi where in his Thalassa he conceives the sexual coitus, "*amphimixie*" of the drives, as the recapitulation of the phylogenetic catastrophes that have marked the evolutionary history of Man. In the genitality are expressed not only the traces memories of the ontogenetic catastrophe (birth), but also those of the phylogenetic catastrophes, which reach an abreaction afterwards. Thus, for Ferenczi, what we call heredity is only the transfer to the descendants of the greater part of the painful task of coping the traumas.

9.5 Identifications as Structural Effects

The second contribution concerns the phenomenon of identification. In analysis, the patient is caught up in the conflictual play of unconscious identifications, constructed since childhood, being made and unmade, sometimes being brutally metamorphosed, without the determinations of these conflicts being entirely intelligible by the internal history of a life. The identifications result from the meeting between the subject and structural determinations (family structure, kinship structure and anthropological structure), local determinations (place in the siblings, sex) and eventual determinations (position in a genealogical history, death, diseases and accidents). The complex conjunction of these determinations hinders any prediction and often any understanding of the causes. But it can be enlightened by the catastrophic interpretation of the genesis of the structures. If we consider that an identification is an effect of structure, in the sense of structuralism where the position in a structure takes precedence over the identity, then the theory of catastrophes brings us a gain of intelligibility. In the so-called "agent" interpretation of catastrophes, we consider the existence of apparent states, which we note here A1 and A2, not in substance as being forms but as virtual positions that can be occupied by agent in a considered system. These positions virtualize each other and are thus identified with positions in a differential structure. The theory of catastrophes becomes a metatheory of structuralism, as shown by the work of Jean Petitot. For example, it is applicable to narrative structures. By interpreting the cusp of the catastrophe of the frontier in the lexicon of linguistics, it becomes an ideal syntactic event informing an elementary narrative structure and distributing places (minima) invested by identity agents (A1 and A2).

At the beginning, there is a single place invested by an agent A1. At the crossing of the first stratum of the cusp, another agent A2 appears, whose influence grows until it comes into conflict with A1. After the crossing of the conflict stratum, A1's influence decreases until it is captured by A2, who remains the only surviving agent. This interpretation models the phenomena of appearance and disappearance of characters and objects in tales, in myths, and more generally in any narrative. The narrative realizes a trajectory on a surface stretched between competing attractors and strewn with catastrophes, thus constituting narrative germs that constitute the poles of Greimas semiotic square.

The theory of catastrophes has also been used in the interpretation of the canonical formula of myths proposed by Lévi-Strauss. This universal formula makes it possible to account for an incomprehensible link between two elements of a myth. This link, for example the identification in an Indian myth between pottery and jealousy, is not deducible from the perceptive experience, and requires a hidden process of attribution of link (transcendental deduction) between these two elements. The internal, hidden process attributes an inverse function (a negative property) to an empirically observable property. But no myth can be explained by an internal semiotic determinism, limited to the interpretation of its content. The determinism of myth I implies the effect of a myth II, absent, possibly unknown, belonging to another people of the same cultural area, which contains actants and functions, which will be integrated in an inverted form into myth I. Why in an inverted form? Because the human mind, fundamentally, builds systems of differences. There can be no symbolic closure by neutralizing differential relations. It is necessary to maintain a circulation of difference so that mythical thought can continue to operate its function. The catastrophic interpretation of the canonical formula, proposed by Jean Petitot, then taken up again, commented on and enriched by Lucien Scubla, allows us to conceive of the existence of catastrophes (dual butterflies) underlying the structure of myths, allowing us to represent (but not to explain) the appearance, disappearance and metamorphosis of the actants, and thus the maintenance of the differential structure. Now, we know since the first works of Lacan, that the identifications between the terms present in the myths are close to those observed in psychoanalysis. The unconscious oedipal identifications, their conflicts, their modifications, which have been described on the structural level, can gain an additional intelligibility by considering them as determined by a catastrophic underpinning. This extra intelligibility consists, for the psychoanalyst, and ideally for the patient, in accepting the fact that the identifications of the subject—correlative of the drama of human existence—are overdetermined by deep, transhistorical dynamics, on which operates the necessity of the symbolic variation, and on which the self, "rider with the untamed mount" to take back the expression of Freud, does not have any means of action, if it is not the half-opened possibility of a fleeting intuition of their powers.

9.6 Conclusions

These three examples do not circumscribe the potential contributions of catastrophe theory to psychoanalysis. Other approaches are possible and have been the subject of other works. A comparative analysis of Freud's Lamarckism and Thom's theoretical biology could be very instructive. Lamarckism, to which Freud was so attached, in spite of the pressing objections of Ernest Jones, consists in accepting the repetition on the scale of an individual life of the events (in this case, the murder of the archaic father inducing guilt and the foundations of the social) which took place on the phylogenetic scale. This myth is invalidated by all contemporary scientific data, but the idea of a transmission of acquired experiences (in a catastrophic form) could take on a new consistency within a conception where the genome is a set of controls acting on a delocalized multidimensional dynamic. The property of internalization of external factors is also strongly suggestive for understanding the constitution of the superego by introjection of parental *imagos*, in other words, in catastrophic language, an external control that has become an internal control by the phenomenon of stabilization of thresholds.

But, in fine, the most consistent contribution of catastrophe theory to psychoanalysis seems to us to be of an epistemological nature. René Thom's invitation to construct a virtual space in which, by immersing a problematic object, we can study its unfolding and trace its generic singularities, must be appreciated as a call to creative thinking. As soon as we wish to take into consideration the extensive globality of psychic life, and for example the construction of mental objects from perception, attention, intentional processes, as soon as we try to understand the fact established by neuropsychology of the modularity of cognitive operations, to make intelligible the neuronal implementation of mental operations (etc.),—all dimensions also implied in the construction of mental objects—all dimensions also involved in psychopathology—then Freudian metapsychology, despite its genius, as well as its structural interpretation by Lacan, become insufficient. It is always possible to decree that being interested in these questions means leaving psychoanalysis, and being satisfied with a local approach—let everyone stay at home and everything will be fine in the best of worlds! But reality ignores disciplinary partitions, and as soon as the clinician takes a taste of the spirit of adventure, attempts a synthesis, and deploys the virtualities of a theoretical imagination capable of welcoming Freudian sexuality and the unconscious as much as neurobiological and cognitive data, then catastrophe theory, which is fundamentally transdisciplinary, offers him a haven. Within the framework of catastrophic models, a vast and flexible conceptual universe, there is indeed room for elaborations where cognition and unconsciousness can be arranged in constructions, certainly virtual, but effective for thought. It will remain for the researcher to verify that the deployments of his catastrophic inspiration remain coherent and provide a gain of intelligibility on the object of study that he has given himself.

Notes For an in-depth approach to the relationship between psychoanalysis and catastrophe theory, see our book Catastrophes of the unconscious, Les éditions Baghera, 2019 and the texts (pdf) gathered under the heading Psychopathology and complexity on the site www.benoitvirole.com.

Bibliography

1. Aulagnier, P.: La violence de l'interprétation, du pictogramme à l'énoncé, Paris, Puf (1975)
2. Derrida, J.: De la grammatologie, Les éditions de Minuit (1967)
3. Ferenczi, S.: Psychanalyse, Oeuvres complètes, Payot (1970)
4. Freud, S.: Oeuvres complètes, psychanalyse, vol. I–XX, Puf 1886–1939.
5. Greimas, A.J.: Sémantique structurale, Puf (1970)
6. Jones, E.: La vie et l'oeuvre de Sigmund Freud, 1953, trois volumes, Puf (1958)
7. Lacan, J.: Le mythe individuel du névrosé, 1952, Seuil (2007)
8. Lacan, J.: Écrits, Le Seuil (1966)
9. Lamarck, Philosophie zoologique, 1809, Garnier Flammarion (1994)
10. Laplanche, J.: Nouveaux fondements pour la psychanalyse, PUF (1987)
11. Laplanche, J., Pontalis, J.-B.: The Language of Psycho-Analysis, translated by Donald Nicholson-Smithn W.W Norton & Company, New York (London, 1967) (1973)
12. Lévi-Strauss, C.: Mythologiques, vol. 4, Plon, 1971 (2009)
13. Petitot, J.: Physique du sens, Éditions du CNRS (1992)
14. Porte, M.: La dynamique qualitative en psychanalyse, Puf (1994)
15. Scubla, L.: Lire Lévi-Strauss, Éditions Odile Jacon (1998)
16. Tesnière, L.: Éléments de syntaxe structurale. Klincksiek, Paris (1982)
17. Thom, R.: Esquisse d'une sémiophysique, Physique aristotélicienne et théorie des catastrophes. InterEditions, Paris (1988)
18. Thom, R.: Paraboles and catastrophes, 1980, Flammarion (1983)
19. Thom, R.: Apologie du logos, Hachette (1990)
20. Thom, R.: Modèles mathématiques de la morphogenèse, 1966, Christian Bourgeois éditeurs (1980)
21. Thom, R.: Prédire n'est pas expliquer, Interview avec Emile Noel, Eshel (1991)
22. Thom, R.: Stabilité structurelle et morphogenèse, seconde édition. Interéditions, Paris (1977)
23. Virole, B.: Psychologie de la surdité, Deboeck université, 1996, troisième édition (2006)
24. Virole, B.: Catastrophes de l'inconscient, Les Editions Baghera (2019)

Chapter 10
René Thom and His Impact on Modeling in Biology

Jacques Demongeot

10.1 Introduction

I met René Thom in 1976 at EPCM laboratory (Physico-Chemical and Molecular Enzymology Lab), where I was preparing my medical thesis supervised in Orsay by François Seydoux and by Jeanine Yon, whose husband Théo Kan, a physiologist from the Institute of Physico-Chemical Biology (IBPC) founded by René Wurmser, knew René Thom and attended his famous Seminar at IHES (Institut des Hautes Etudes Scientifiques at Bures-sur-Yvette). I had discovered IHES during a visit to Claude-Paul Bruter in 1977, to which I showed my medical thesis manuscript. René Thom gave a lecture that year at Orsay, during which I risked a few insignificant remarks and he glared at me, saying: "You must be one of those young scholars who rush to the computer before thinking!" This made a big impression on me, and two years later I rushed to the Ecole Normale Supérieure for the first meeting on Theoretical Biology which he organized, with Michel Thellier and Pierre Delattre, and which they edited with the publisher Maloine. Then, for 25 years, we (his medical students Yannick Kergosien, Etienne Labeyrie and Jacques Viret, his colleagues and friends Claude-Paul Bruter, Yves Bouligand, and many others, etc.) were among his unconditional "groupies"... He always encouraged me, in particular when we published for the first time with Editions du CNRS the Acts of the meetings of our national Society of Theoretical Biology in France (AMTB, then SFBT), then at European level (ESMTB), and he participated in all the scientific meetings that we organized, even if they were dedicated for example to respiratory physiology. In a subtle way, he traced scientific leads that we faithfully followed and which turned out to be in fact real programs. I spent, alongside René Thom and Yannick Kergosien, six unforgettable sabbatical months in 1985 at IHES, which partly nourished me scientifically over the next 30 years, the other part being attached rather to the magisterium of his

---

J. Demongeot (✉)
Grenoble Alpes University, Grenoble, France
e-mail: jacques.demongeot@yahoo.fr

© The Author(s), under exclusive license to Springer Nature Switzerland AG 2024
I. Marcos and C. Morier (eds.), *The Relevance of René Thom*, Lecture Notes in Morphogenesis, https://doi.org/10.1007/978-3-031-54982-3_10

great friends from Poitiers, Pierre Gavaudan, Marcel-Paul Schützenberger and their disciple, my Master Jacques Besson.

10.2 Historical Overview of Biological Ideas and Models

René Thom, after his Fields medal obtained in 1958, returned almost naturally to the humanities and biology, where he excelled in high school. He claimed, in a very prosaic way, to have done mathematics, because "at the time when the war was approaching, it was the general opinion that it was better to do mathematics to enter the artillery than to go into the infantry where one was condemned to certain death...", but he attentively followed the extraordinary condensation of biological ideas that occurred in 1952–1953:

- Free Monoids and Linguistics [9, 10, 61]
- Evolution and Amino Acids [44]
- Cellular Automata [68, 72]
- Epigenetics of birds [73–75]
- Structure of DNA [76].

There had been no other example of a quasi-simultaneous irruption of new concepts in various scientific fields, ranging from biology to mathematics, via linguistics, since the fertile period of the Age of Enlightenment, from 1750 to 1765, with the following philomats cited in reference in various fields:

- Epidemiology [3]
- Linguistics [14]
- Encyclopedia [29, 30]
- Principle of Least Action [32, 46]
- Theory of the inner mold [42]
- Botanical cladistics [71]
- Mechanistic physiology [49, 50]
- Physiocratic econometrics [15, 54]
- Experimental physiology [16].

The great classical culture of René Thom, from Heraclitus to Waddington (whom he considered his Master in biology) led him to be one of the major players in the last great condensation of ideas, from 1971 to 1974:

- Topology and Perception [4]
- Self-organization of the living systems [2, 31]
- Dissipative systems [40]
- Bifurcations and chaos [36]
- Origin of Life [52]
- Child psychology [55]
- Strange attractors [59]

- Ergodicity [62]
- Catastrophe theory [64].

There is a remarkable continuity between the ideas of the scholars cited above. For example, the notion of attractor is already present in Buffon (Fig. 10.1), for whom the perturbation of the "inner mold" is damped, to allow the constancy of the organ morphology and function, prefiguring the organ-function relationship of Lamarck's and Claude Bernard's homeostasis. This notion of attractor is then defined by many authors [12, 41, 45, 59, 64, 77] and now constitutes a notion very used by biologists, especially embryologists [56] or therapists. In the latter case, we can consider the state of good health as being a physiological attractor, which supports ordinary perturbations (due to food, breathing, variation in ambient temperature, etc.), dampened by the mechanisms of homeostasis (type 1 perturbation on Fig. 10.1).

When the perturbation amplitude exceeds an acceptable level, it can create pathological behavior (possibly lethal) and the organism places itself, after a possible catastrophe-type jump, on a temporary attractor (high fever, sweating, shallow breathing, etc.), that an appropriate therapeutic counter-disturbance will return to

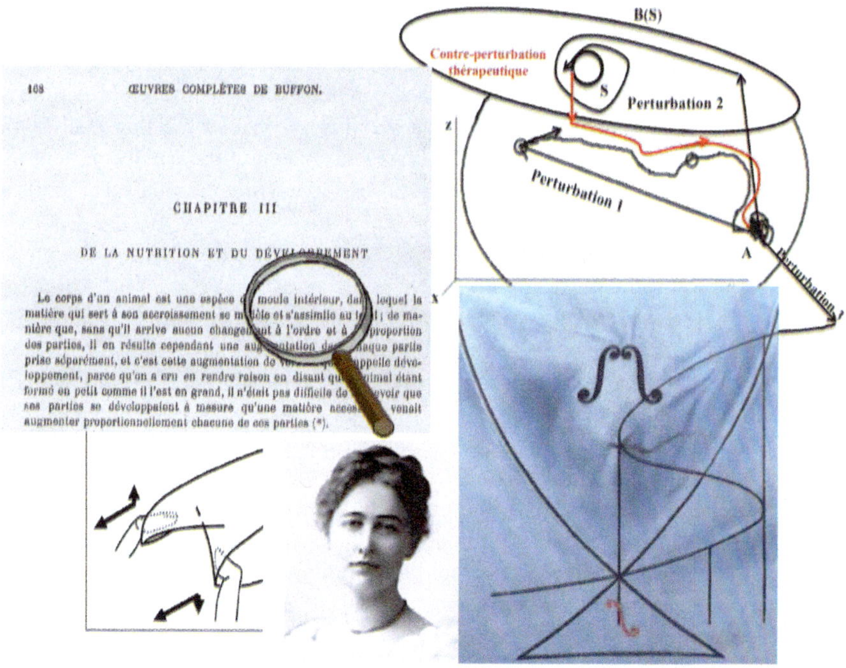

Fig. 10.1 Top left: Buffon's inner mold theory. Top right, perturbations of a physiological attractor. Bottom left, birth of a gather. The fingers are used for the first time by J. Demongeot, according to [46]. Bottom middle, Maud Menten, co-author of the Michaelis–Menten model, used for tree growth. At the bottom right, S. Dali places his mustache instead of his fingers, in a painting entitled "The Dovetail", series of Catastrophes

the borders of the basin of attraction of the physiological attractor (type 2 perturbation). If the perturbation is too strong, the organism can escape into a lethal behavior (type 3 perturbation), with an infinite accumulation of input variables (for example glucose, which is no longer metabolized). This medical reading of the effect of a perturbation is only the pictorial translation of mathematical concepts, such as the stability of a dynamical system, on which René Thom reflected a great deal [5–7, 19–22, 64]. To define this stability, we follow an article published in the Bulletin du Groupe d'Etude des Rythmes Biologiques [17], in which a dynamical system can present several types of stability:

- the weak asymptotic stability corresponds to an independence with regard to the initial conditions: whatever the initial condition x, the system as time tends to infinity (circumstance called asymptotic) towards a set of recurrent states denoted L(x) contained in a set called attractor A, if x belongs to a subset, called attraction basin of A and denoted B(A) (Fig. 10.1). By definition, the attractor $A = L(B(A)) = U_{x \in B(A)} L(x)$, that is verifies a invariant property for the composed operator LoB [13]. The weak asymptotic instability corresponds for example to the absence of an attraction basin: this is the case for the harmonic pendulum, whose asymptotic behavior depends completely on the initial conditions, each trajectory of the system does not "attract" no other; the only stability observed in this case, called Lyapunov stability, corresponds to the fact that a perturbation of the initial conditions is never increased when time tends to infinity. In the case of a strange attractor, weak stability is verified; on the other hand, there is no strong asymptotic stability, that is to say that initial conditions tending to merge will not lead to the cancellation of the asymptotic phase shift of the trajectories,
- the strong asymptotic stability corresponds to a good resistance to the noise on the state variables, if the noise confines them in the attraction basins. This resistance gave rise to the concept of π-confiner, which is none other, in a noisy system, than the subset of the state space in which the system remains asymptotically present with the probability π, if the initial conditions remain inside the same subset, called the π-confinement basin [23],
- the structural stability corresponds to robustness with respect to the parameters: when these vary, the number of attractors and the relationships between their basins are invariant; a value of the parameters causing a change in this number or in this topography is called bifurcation value of the system, which is then said to be structurally unstable. The structural stability corresponds to the resistance to the noise of system parameters, when this noise leaves their bifurcation values out of reach.

10.3 Example of Tree Growth Modeling

René Thom described the growth of trees as a "surge towards the light", maximizing the capture of solar energy by photosynthesis. A very simple model, valid for the growth of *Araucaria* tree (Fig. 10.2 top left) can account for the main stages observed in its morphogenesis. The local apices, at the end of its branches, promote the synthesis, in its meristems, of a plant hormone, auxin, which stimulates local growth and promotes the synthesis of an inhibitor which diffuses, prohibiting the growth of secondary axillary buds. A functional apex therefore inhibits the other buds of its branch. Plant photo-tropism (e.g., heliotropism of sunflowers) is due to the fact that auxin is more photo-oxidized on the lit wall of the plant, causing it to migrate towards the wall on the unlit side (Fig. 10.3 below), which then experienced stronger growth.

The modeling of this growth uses the variables X_1, cell mass of the apex, and X_2, mass of the primary axillary buds, which succeed it on the branch.

The differential system ruling the dynamics of X_1 and X_2 can be written as [11]:

$$\frac{dX_1}{dt} = s_1X_1 + \frac{aX_1}{K_1 + X_1 + X_2} - d_1X_1, \; \frac{dX_2}{dt} = s_2X_2 + \frac{aX_2}{K_2 + X_1 + X_2} - d_2X_2$$

The second member of the equations uses the Michaelis–Menten kinetics with competitive inhibition, whose maximal velocity a is proportional to the auxin's concentration. Ki's are constants modulating this velocity at weak concentrations, and s_i and $\mathrm{d_i}$ are respectively input and ouput constants of X_i.

The stationary state equations are:

$$(d_1 - s_1)X_1^2 - (s_1K_1 + a - d_1K_1)X_1 + (d_1 - s_1)X_1X_2 = 0$$

$$(d_2 - s_2)X_2^2 - (s_2K_2 + a - d_2K_2)X_2 + (d_2 - s_2)X_1X_2 = 0$$

Let suppose that $K_1 = K_2, s_1 = s_2 = \mathrm{s} \;\mathrm{et}\; d_1 = d_2 = d$, then we obtain by subtracting the second equation to the first: $(d - s)(X_1 + X_2 + K) - a = 0$, from which we get the values of the two stable stationary X^* et X^{**}:

$$X_1^* = \frac{(s - d)K + a}{d - s}, X_2^* = 0; X_1^{**} = 0, X_2^{**} = \frac{(s - d)K + a}{d - s}$$

We can prove that even the kinetics is highly non-linear of Hill type:

$$\frac{aX_i^n}{K + \sum_{k=1}^{m} X_k^n}$$

where n is the cooperativity parameter, then the system is of gradient type thanks to a change of variables: $x_i = (X_i)^{1/2}$. Its potential P is defined by:

Fig. 10.2 Top left: *Araucaria* tree. Top right: Growth of *Rosa chinensis viridiflora* from Goethe's garden, photographed in Jacques Besson's garden. Middle: Potential surface *P* and location of its minima. Bottom: Potential gradient dynamics *P* (left) and Lyapunov dynamics (right), where the energy *P* decreases on the trajectories, but without its gradient vector being equal to the opposite of velocity

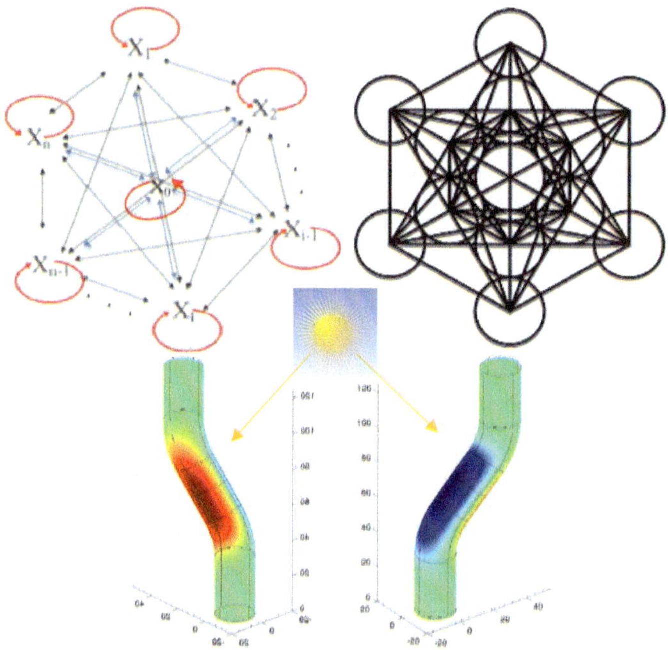

Fig. 10.3 Top left, Jacobian graph of the differential growth system of a tree. Top right, superposition of the projections of the Platonic solids (diagram of Metatron). Below, simulation of the accumulation of auxin (in red) on the wall opposite the light [33–35]

$$P(x) = \frac{\sum_{k=1}^{m}(d_k - s_k)x_k^2}{4} - \frac{aLog\left(K + \sum_{k=1}^{m} X_k^{2n}\right)}{4n}$$

and the gradient system ruling the x_i's is given by the equations:

$$\frac{dx_1}{dt} = \frac{-\partial P}{\partial x_1}, \frac{dx_2}{dt} = \frac{-\partial P}{\partial x_2}$$

Figure 10.2 middle shows the potential P and its two minima, which are the two stationary states of the system, as well as their basin of attraction. Figure 10.3 shows the Jacobian graph of the system, when the number of potential buds is equal to 6. This graph is completely connected and all its edges are negative, except the loops. When the number of buds is equal to 10, the graph is the projection of the superposition of the solids of Plato's Timaeus. The biological interpretation of these results is as follows: when the apical cells develop, the auxin they secrete inhibits the axillary buds. The development of the branch bearing the apex makes it more difficult for auxin to diffuse over an increasing length, which then releases the growth of the nearest axillary buds, which have reached, thanks to the growth of the branch, in contact with the border of the phototropic zone. These primary axillary buds in turn inhibit the secondary axillary buds that follow them, which will themselves only be activated when the critical distance for auxin diffusion from the first axillary buds is again reached. An example of the shape thus obtained for the adult tree is that of the *Araucaria* tree (Fig. 10.2 top left). If the parameter a follows a slow dynamic, it can cause a change in the attraction basin landscape, provoking for example that the system crosses the border represented by the crest of the surface representative of the potential P on Fig. 10.2 middle, causing a catastrophe, which may correspond to the hormonal change giving birth of new branches in *Araucaria*. The slow dynamics makes it possible to move on an isocline surface of the system, cubic in case of Hill kinetics of order 2. We can then observe a swallow tail type catastrophe (Fig. 10.1 bottom left), allowing to jump suddenly from one sheet to another of the isocline surface.

In his famous book *Metamorphoses*, Ovide declares: "Before the sea, the earth and the sky that covers everything, nature in the whole universe presented only one aspect, which was called Chaos. It was a coarse and confused mass, nothing but an inert mass, a piling up of seeds of things, of divided and badly joined elements" [49, 53]. In Ovidian chaos, each element is disconnected from the others, the contrary of a fully connected newborn brain. In Ovidian chaos three independent elements cannot evolve and remain in the same initial state. Ovid goes on to say: "A god, of a more willing nature…, freeing these things from the blind mass, then fixed them in place, separately, in peace and interacting harmony". By introducing interactions in a fully connected system of 3 Boolean elements (Fig. 10.4), with 50% of positive interaction (activation) and 50% of negative ones (inhibition) and by choosing by chance the direction of the interaction supposed to be unidirectional between elements, we observe different dynamical behaviors as shown on the Fig. 10.4.

On the simple example of Fig. 10.4, we see that the number and nature of the attractors of a Boolean dynamics [43], where the state of an element is fixed to 1 (resp. 0) if it is activated (resp. inhibited) by another in state 1, only depend on the number and nature of the circuits in the interaction graph (a circuit being a path of the graph starting on a node and returning to it passing one and only on time through the intermediate nodes).

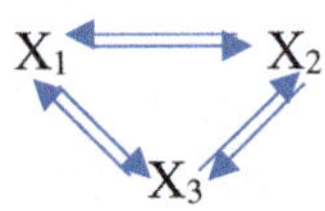

Nb circuits	Nb inhibitions	Attractors
0	0, 1, 2 or 3	1 fixed point depending on initial conditions
1 positive	0	2 fixed points: 11111… and 00000…
1 negative	1	1 limit cycle: 111000111000111000…
1 positive	2	2 fixed points: 100100100… and 011011011…
1 negative	3	1 limit cycle: 10101010101010…

Fig. 10.4 Top: Interaction graph of a system with 3 elements. Bottom: Attractors of the system whose number and nature depend on the number and nature of the circuits in the interaction graph

The type of growth by apparent successive metamorphoses of buds giving leaves, then flowers was illustrated by Goethe in his book *Die Metamorphose der Pflanzen* [70], who observed, in his Frankfurt garden, a green rose imported from China in 1743 of the *Rosa chinensis viridiflora* type, where all the elements of the plant (stem, leaf, petals and sepals) are green (a phenomenon called phyllody), each seeming to result from the continuous metamorphosis of the preceding element (Fig. 10.2 top right). These elements belong to a floral regulatory network (Fig. 10.5 top), which has been studied in *Arabidopsis thaliana* [25], by deleting various combinations of floral identity genes and examining what the resulting organs looked like (Fig. 10.5 bottom left). The simulation of a Boolean model of the network the expression of floral genes then found the experimental botanical classification of *Rosa viridiflora* as a mutant flower (Fig. 10.5 bottom right). The explanation sought by Goethe for the flower color was finally found.

10.4 Embryogenesis Modeling

René Thom after his Masters in biology Thompson D'Arcy [67] and Waddington [73–75] devoted part of his theoretical morphogenetic studies to embryology. A first example in this field is provided by La Mettrie [50, 51], this Enlightenment philosopher who said "Do you know why I still pay attention to men? It's because I seriously believe them to be machines…", and which stated, in his book *Man more than Machine* (1748): "All the curious have seen the most skillful of these animalcules insinuate themselves into the egg that provides him with the woman" (Fig. 10.5 left), a prefiguration of what shows the simulations of the model by Dupont-Goldbeter [39] on the irreversible asymptotic structuring of the spatial calcium waves after the fertilization (analogous to the tubulin waves observed in Glade et al. [37, 38] inside ovum, early after the striking of the pellucid membrane of the ovum by the cap of the first spermatozoon, "the most skillful" (Fig. 10.6).

The fertilization of the ovum by the spermatozoon causes the first break in spherical symmetry of the embryo by establishing a cephalo-caudal calcium gradient (first metameric organization of the embryo in a calcium periodic structure, ancestor of the vertebral column), then a second rupture appears by breaking the symmetry of revolution: after the skeleton, it is indeed imperative to build a tube which will become the ancestor of the digestive system. This stage is called the gastrulation (Fig. 10.7), and a model of the same type as that used for tree growth can be used to describe the different steps of its dynamics.

The gastrulation model uses as a basic law the cell division law [35], stated by René Thom at the Darboux amphitheater of the Henri Poincaré Institute in 1992, which can be summarized by saying that a cell divides when its outer surface (its perimeter in two dimensions), which receives energy and nutrients, related to its internal volume (its area in two dimensions), where they are metabolized, falls below a critical threshold (Fig. 10.7 below). The modeling is discrete and processes all the cells of the embryonic structure in parallel. The simulation makes it possible to

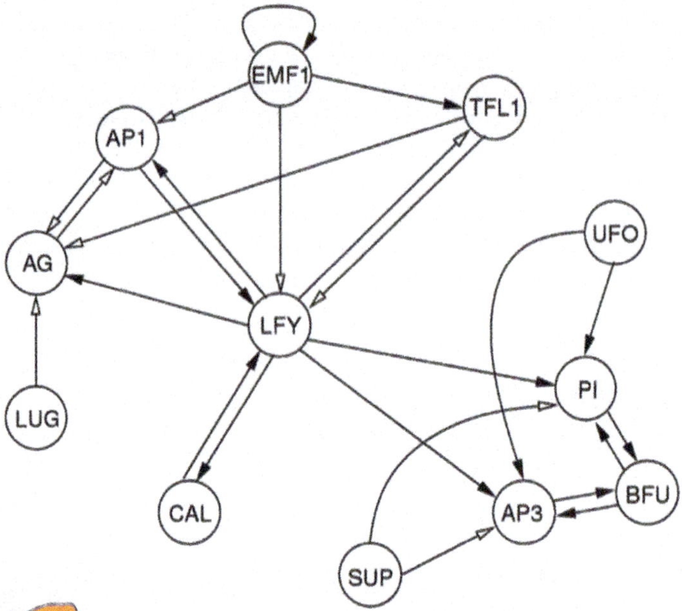

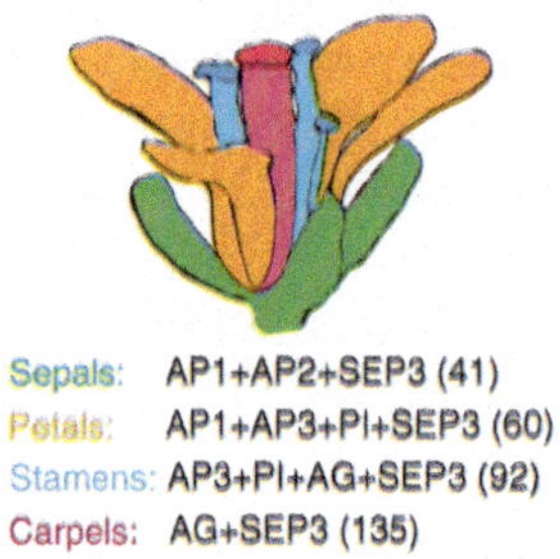

Attractor type	Gene order: EMF1, TFL1, LFY, AP1, CAL, LUG, UFO, BFU, AG, AP3, PI, SUP, AP2	Cell type
Fixed point 1	0001000000001	Sepal
Fixed point 2	0001000101100	Petal
Fixed point 3	0000000010000	Carpel
Fixed point 4	0000000111100	Stamen
Fixed point 5	1100000000000	No flower
Fixed point 6	1100000101100	Mutant flower

Fig. 10.5 Top: Floral regulation network. Activation arrow tips are in black and inhibition ones in white. Bottom left: Differentiation of floral cell types. Bottom right: Gene expression of the floral regulatory network and cell types expressed

show in section the constitution of a primordial gastric sulcus, corresponding to the experimental observations (Fig. 10.7 middle) by Eric Wieschaus's team [63], without involving morphogen, but only by the obligation made to the cell walls of keep in touch during divisions imposed by the Thom's law.

10.5 Beyond Modeling, the Philosophy of Form

René Thom says, quoting Aristotle, that "a being has an edge, it is separated from the surrounding space. In short, the edge of the thing is its form" [18, 65, 66]. The description of this frontier can result from the simulation of a mathematical

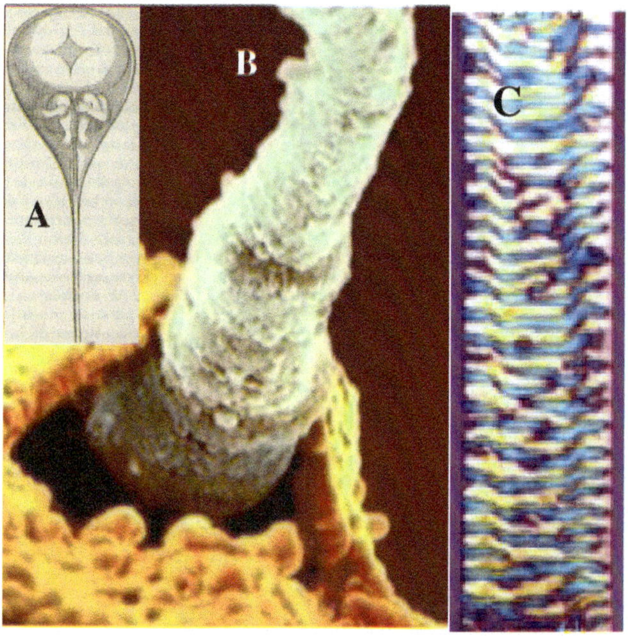

Fig. 10.6 **A**: Homunculus de Hartsoeker. **B**: Spermatozoon striking the pellucide membrane. **C**: Structures de microtubules in a spectrophotometer cell. In presence of gravity, tubulin waves self-organize spontaneously [1]

model or come from an artistic representation whose underlying knowledge can be very profound. For example, unable to state directly in a treatise, like Vesalius, his anatomical knowledge, Michelangelo lets them guess in his David, where the position of the muscles and tendons is perfect. Similarly, examination of the fresco on the Creation of Man on the ceiling of the Sistine Chapel inevitably brings to mind a sagittal section of the human brain (Fig. 10.8), placed on the drawing of God stretching out the hand to his creature. The two fingers that touch prefigure a first synapse. The angels surrounding God correspond to known structures: the leg of the angel supporting God corresponds to the hypophysis, his chest to the piriform cortex, the bottom of his face to the olfactory bulb and the top to the orbito-frontal cortex. The hair of the lower angel is at the level of the amygdala and the two angels are separated by the Sylvius fissure.

More than a simple anatomical paraphrase, Michelangelo's fresco is a functional metaphor and both historians and neuroscientists are lost in conjectures about the origin of the knowledge Michelangelo had of the anatomy and functions of the human brain (https://www.nationalgeographic.fr/histoire/la-chapelle-sixtine-le-chef-doeuvre-de-michel-ange). From comparable examples, Thom was able to extract a universal theory of the genesis of forms, less naturalistic, but more generic

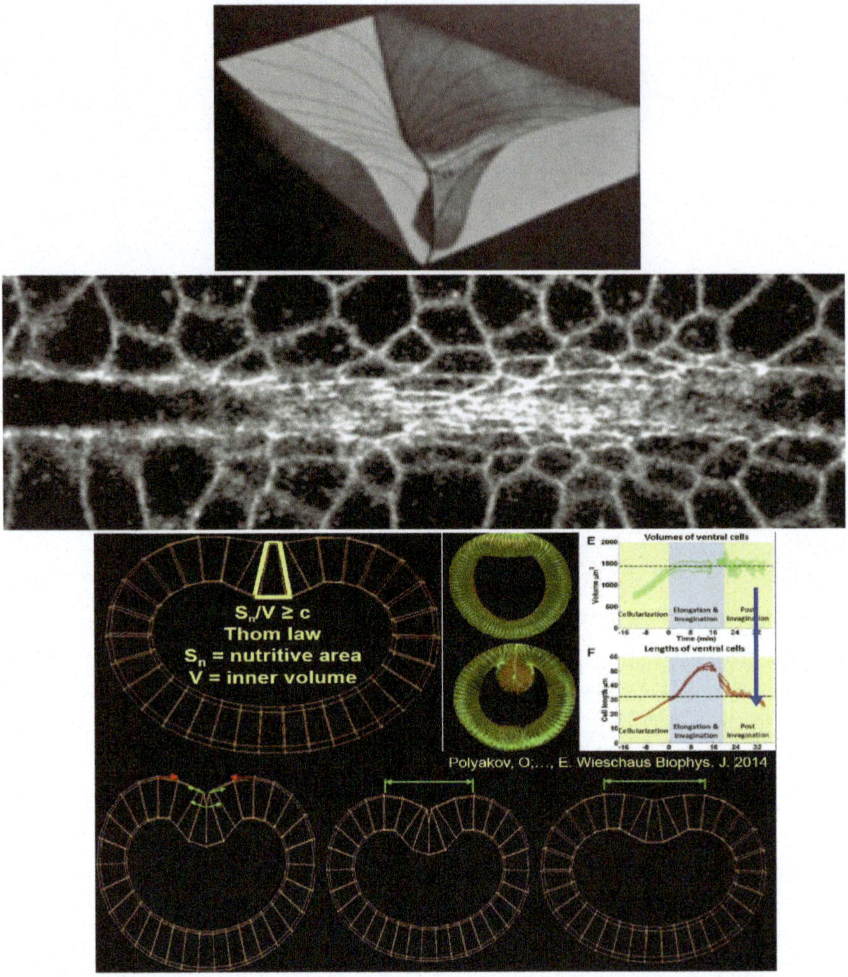

Fig. 10.7 Top left: a plaster model of Needham representing embryonic development [48]. Middle: opening of the digestive tract during gastrulation (experiment by Wieschaus [56]). Bottom: gastrulation simulation using the Thom's growth law

than that of d'Arcy Thompson, because applicable to very abstract, linguistic or mathematical structures. Thanks to his great theory of cobordism, René Thom received the Fields Medal in 1958, at the age of 34.

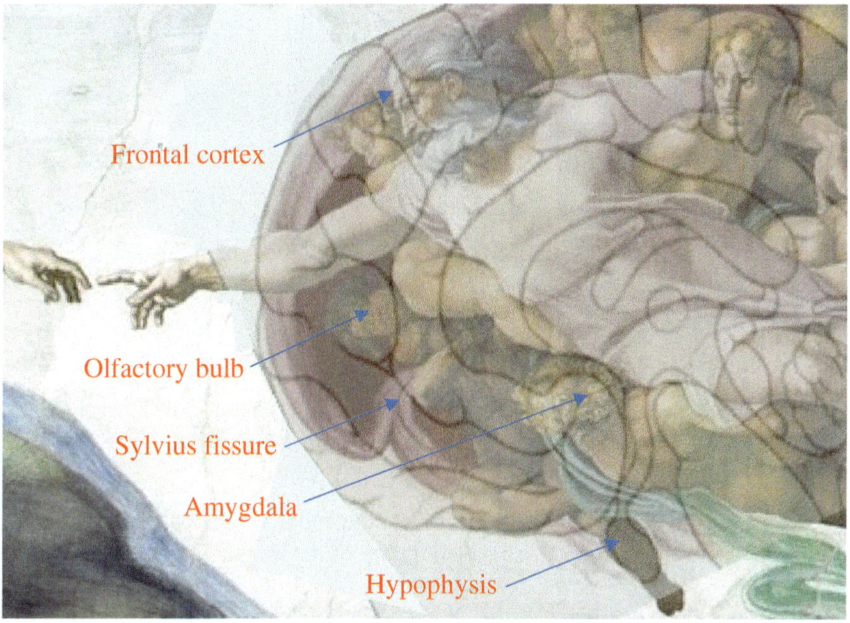

Fig. 10.8 Representation of God creating Adam, with superimposing a sagittal section of the human brain showing corresponding anatomic structures

10.6 Conclusion

At one of the first meetings of the French community of theoretical biology, organized at the Ecole Normale Supérieure in 1978 by Delattre and Thellier [18], to conclude a debate on the definition of a biological model, Boris Rybak [60] went up to the chair and, slamming a heavy total knee prosthesis on the desk, declared: "The model is the prothesis!" I still remember the leap René Thom made. At the Poincaré Institute, at a later meeting, Thom declared: "No need for a model, that's obvious!" when asked what experiment would confirm his theory about the length of the intermitotic interval. "Si non e vero, e bene trovato," concluded the crowd.

The history of biological modeling shows that, for 250 years, concepts have preceded equations, and that simulations have only confirmed experimental observations, themselves sometimes preceded by rational speculations.

Whether or not the work of the modeler was then completed, part of the explanatory work has always been carried out by the initial, formidable and indispensable metaphor that Francesco Redi illustrated so well [58]: placing a piece of meat in a box comprising narrow openings, he materialized, at the level of these slits, the functional intermediary which he suspected—the laying of flies—thus ruining the hypothesis of the spontaneous generation of worms which seemed to appear without cause on any dead animal. The experimental set-up was both the end of an implacable logical funnel unmasking the existence of the eggs (which the accumulation

Fig. 10.9 From left to right, Jullien Offray de la Mettrie, René Thom and Carl-Gustav Jung, three opposite philosophies of life, but the same originality in the development of scientific concepts

around the slits made visible), and a concrete metaphor for the path linking the fly to the worm, the implied comparison being that of the animal reproductive process (Fig. 10.6).

For Mendel, the visible and the hidden—the appearance and the being—the phenotype and the genotype, constituted a metaphor as long as the genes were not discovered. For Claude Bernard, the distinction between "germ", the teleological support of the vitalist program of homeostasis and the physicochemical "machinery" responsible for the execution of this program is a metaphor all the stronger because it partly appealed to biological vocabulary. Many other examples prove the power of this metaphorical approach, prolegomena to modelling, in the first attempts to explain the observed phenomena, sometimes coming from distant scientific fields, such as linguistics in Thom, who assimilates the endoderm to the subject, the mesoderm to the verb and the ectoderm to the object, thus summarizing the major functions devolved at the adult stage to these three types of embryonic tissues, or coming from mathematics in Misha Gromov, who sees in the congruence between differentiable varieties linked to the three-dimensional structure of proteins and nucleic acids, the stereochemical origin of life [8].

In final conclusion, we will quote Jacques Viret, who, in his last work on an imaginary dialogue between his Master René Thom and his philosophical model Carl-Gustav Jung (Fig. 10.9), tells us: "Thom and Jung knew first of all to observe with great acuity, that is to say to examine with a curious and critical mind, firstly the relationships that certain things have between them, secondly men and the relationships between them they maintain, but whether it be, things or people, what counted for them was to clarify, classify, order, then understand the different abstract "forms" of these relations", in short to model the biological or psychological nature [24–29, 54, 59, 69], in an almost identical approach, which constitutes, since the Age of Enlightenment, the heart of scientific thought.

References

1. Abbas, L., Demongeot, J., Glade, N.: Synchrony in reaction-diffusion models of morphogenesis: applications to curvature-dependent proliferation and zero-diffusion front waves. Phil. Trans. R. Soc. A **367**, 4829–4862 (2009)
2. Atlan, H.: L'Organisation biologique et la Théorie de l'information. Hermann, Paris (1972)
3. Bernoulli, D.: Essai d'une nouvelle analyse de la mortalité causée par la petite vérole et des avantages de l'inoculation pour la prévenir, pp. 1–45. Mémoires de Mathématiques et de Physique de l'Académie Royale des Sciences (1760)
4. Bruter, C.-P.: Topologie et Perception, tome 1, Bases philosophiques et mathématiques, Préface de R. Thom. Maloine, Paris (1974)
5. Bruter, C.-P.: Topologie et Perception, tome 2, Aspects neurophysiologiques. Maloine, Paris (1976)
6. Bruter, C.-P.: Topologie et Perception, tome 3, Considérations socio-psychologiques et linguistiques. Maloine, Paris (1986)
7. Bruter, C.-P.: Les Architectures du Feu: Considérations sur les Modèles. Flammarion, Paris (1982)
8. Carbone, A., Gromov, M.: Functional labels and syntactic entropy on DNA strings and proteins. IHES/M/01/54, IHES, Bures/Yvette (2001)
9. Chomsky, N., Schützenberger, M.-P.: The algebraic theory of context-free languages. In: Computer Programming and Formal Systems. North Holland, Amsterdam (1963)
10. Chomsky, N., Piaget, J.: Théories du langage, Théories de l'apprentissage. Débat au Centre Royaumont. Editions du Seuil, Paris (2018)
11. Cinquin, O., Demongeot, J.: Positive and negative feedback: striking a balance between necessary antagonists. J. Theor. Biol. **216**, 229–241 (2002)
12. Cosnard, M., Demongeot, J.: On the definitions of attractors. Lect. Notes Math. **1163**, 23–31 (1985)
13. Cosnard, M., Demongeot, J.: Attracteurs: une approche déterministe. Comptes Rendus de l'Académie des Sciences, Mathématiques Série I **300**, 551–556 (1985)
14. De Brosses, C.: Traité de la formation mécanique des langues. Saillant, Paris (1765)
15. de Caritat, J-.A.N.: Du calcul integral. Didot, Paris (1765)
16. De Montbesse, J.: La modélisation et ses limites en biologie. Bull. SFBT **17**, 9–20 (1992)
17. Dalleur, P.: Fécondité de la notion de «bord» des formes vivantes chez Thom. Revue Philosophique de Louvain **104**, 312–346 (2006)
18. Delattre, P., Thellier, M.: Elaboration et Justification des Modèles, tomes 1 & 2. Maloine, Paris (1979)
19. Demongeot, J.: Existence de solutions périodiques pour une classe de systèmes différentiels gouvernant la cinétique de chaînes enzymatiques oscillantes. Lect. Notes Biomath. **41**, 40–62 (1981)
20. Demongeot, J., Besson, J.: Code génétique et codes à enchaînement I, C.R. Acad. Sc. Série III **296**, 807–810 (1983)
21. Demongeot, J., Laurent, M.: Sigmoidicity in allosteric models. Math. Biosci. **67**, 1–17 (1983)
22. Demongeot, J., Jacob, C.: "Confinors: a stochastic approach", C.R. Acad. Sc. Série I **309**, 699–702 (1989)
23. Demongeot, J., Glade, N., Forest, L.: Liénard systems and potential-Hamiltonian decomposition. Methodology. Comptes Rendus Mathématique **344**, 121–126 (2007)
24. Demongeot, J., Goles, E., Morvan, M., Noual, M., Sené, S.: Attraction basins as gauges of robustness against boundary conditions in biological complex systems. PLoS ONE **5**, e11793 (2010)
25. Demongeot, J., Hazgui, H.: The Poitiers school of mathematical and theoretical biology: Besson-Gavaudan-Schützenberger's conjectures on genetic code and RNA structures. Acta Biotheor. **64**, 403–426 (2016)
26. Demongeot, J., Hazgui, H.: Entropy, a tool for system biology applied to the genetic regulatory networks. LINKs **2**, 128–144 (2019)

27. Demongeot, J., Jelassi, M., Taramasco, C.: Big data approach for managing the information from genomics, proteomics, and wireless sensing in e-Health. In: Big Data and Remote Sensing: Acquisition, Visualisation and Interpretation, pp. 1–37. Springer, New York (2019)
28. Demongeot, J.: The will of nature and the will of man. In: Quantum Psyche II, éd., pp. 1–16. Carminati, Francisco, Giugi Editions, Genève (2020)
29. Diderot, D., Le Rond d'Alembert, J.: L'Encyclopédie, chez Briasson, David, Le Breton. Durand, Paris (1751)
30. De Réaumur (Ferchault), R.-A.: Observations sur la digestion des Oiseaux. Journal des sçavans, juin 164–177 (1753)
31. Eigen, M.: Selforganization of matter and the evolution of biological macromolecules. Naturwissenschaften **58**, 465–523 (1971)
32. Euler, L.: Sur le principe de la moindre action, vol. 7. Mémoires de l'académie des sciences de Berlin, pp. 199–218 (1753)
33. Forest, L., San Martin, J., Padilla, F., Chassat, F., Giroud, F., Demongeot, J.: Morphogenetic processes: application to cambial growth dynamics. Acta Biotheor. **52**, 415–438 (2004)
34. Forest, L., Padilla, F., Martínez, S., Demongeot, J., San Martín, J.: Modelling of auxin transport affected by gravity and differential radial growth. J. Theor. Biol. **241**, 241–251 (2006)
35. Forest, L., Demongeot, J.: Cellular modelling of secondary radial growth in conifer trees: application to Pinus radiata. Bull. Math. Biol. **68**, 753–784 (2006)
36. Gavrilov, N., Shil'nikov, L.: On three-dimensional dynamical systems close to systems with a structurally unstable homoclinic curve. Mathematicheskii USSR-Sbornik **17**, 467–485 (1972)
37. Glade, N., Demongeot, J., Tabony, J.: Numerical Simulations of microtubule self-organisation by reaction and diffusion. Acta Biotheor. **50**, 232–248 (2002)
38. Glade, N., Demongeot, J., Tabony, J.: Microtubule self-organisation by reaction-diffusion processes causes collective transport and organisation of cellular particles. BMC Cell Biol. **5**, 23 (2004)
39. Goldbeter, A., Dupont, G., Berridge, M.J.: Minimal model for signal-induced Ca2+ oscillations and for their frequency encoding through protein phosphorylation. Proc. Natl. Acad. Sci. USA **87**, 1461–1465 (1990)
40. Glansdorff, P., Prigogine, I.: Structure, Stabilité et Fluctuations. Masson, Paris (1971)
41. Haraux, A.: Attractors of asymptotically compact processes and applications to nonlinear partial differential equations. Commun. Part. Differ. Equ. **13**, 1383–1414. https://www.nationalgeographic.fr/histoire/la-chapelle-sixtine-le-chef-doeuvre-de-michel-ange (1988)
42. Leclerc de Buffon, G.-L.: Histoire naturelle, générale et particulière, avec la description du Cabinet du Roy. Imprimerie Royale, Paris (1749)
43. LeCun, Y.: Learning processes in an asymmetric threshold network. In: Disordered Systems and Biological Organization, pp. 233–240. North Holland, Amsterdam (1986)
44. Miller, S.: A production of amino acids under possible primitive earth conditions. Science **117**, 528–529 (1953)
45. Milnor, J.: On the concept of attractor. Commun. Math. Phys. **99**, 177–195 (1985)
46. Moreau de Maupertuis, P.-L.: Essai de Cosmologie. Pierre Mortier, Amsterdam (1750)
47. Morier, C.: La figure comme forme d'un processus évolutif, ou l'apport de René Thom. In: Figure–du mot au concept, pp. 85–111. Presses Universitaires de Grenoble (PUG), Grenoble (2015)
48. Needham, J.: Order and Life. Cambridge University Press, London (1936)
49. Nisard, D.: Ovide. Garnier, Paris (1850)
50. Offray de la Mettrie, J.: L'homme plus que Machine. Payot, Genève (1748)
51. Offray de la Mettrie, J.: Ouvrage de Pénélope ou Machiavel en médecine. Fayard, Paris (1750)
52. Paecht-Horowitz, M.: The origin of Life. Angew. Chem. **12**, 349–356 (1973)
53. Phillippy, P.: Altera Dido': the model of Ovid's Heroides in the poems of Gaspara Stampa and Veronica Franco. Italica **69**, 1–18 (1992)
54. Pham, D.T., Demongeot, J., Baconnier, P., Benchetrit, G.: Simulation of a biological oscillator: the respiratory rhythm. J. Theor. Biol. **103**, 113–132 (1983)
55. Piaget, J., Inhelder, B.: The Psychology of the Child. Basic Books, New York (1972)

56. Polyakov, O., He, B., Swan, M., Shaevitz, J., Kaschube, M., Wieschaus, E.: Passive mechanical forces control cell-shape change during Drosophila ventral furrow formation. Biophys. J. **107**, 998–1010 (2014)
57. Quesnay, F.: Le Tableau économique. Presse des Petits Appartements, Versailles (1758)
58. Redi, F.: Esperienze intorno alla generazione degl'insetti. Carlo Dati, Florence (1668)
59. Ruelle, D., Takens, F.: On the nature of turbulence. Commun. Math. Phys. **20**, 167–192 (1971)
60. Rybak, B.: Du modèle naturel aux modèles artificiels et inversement. In: Elaboration et justification des modèles, pp. 57–67. Applications en biologie. Maloine, Paris, tome 1 (1978)
61. Schützenberger, M.-P.: Le problème des mots dans les treillis modulaires libres. Comptes Rendus Académie des Sciences **237**, 507–508 (1953)
62. Sinaï, Y.G.: Gibbs measures in ergodic theory. Russ. Math. Surv. **166**, 21–69 (1972)
63. Sweeton, D., Parks, S., Costa, M., Wieschaus, E.: Gastrulation in Drosophila: the formation of the ventral furrow and posterior midgut invaginations. Development **112**, 775–789 (1991)
64. Thom, R.: Stabilité structurelle et morphogenèse. InterÉditions, Paris (1972)
65. Thom, R.: Modèles mathématiques de la morphogénèse, Collection 10–18. Union Générale d'Éditions, Paris (1974)
66. Thom, R.: Prédire n'est pas expliquer. Esthel, Paris (1991)
67. Thompson D'Arcy, W.: On Growth and Form. Cambridge University Press, Cambridge, UK (1917)
68. Ulam, S.: Random processes and transformations, vol. II. In: Proceedings of the International Congress of Mathematicians, pp. 264–275. American Mathematical Society, Providence, R. I (1952)
69. Viret, J.: Thom et Jung, un dialogue imaginaire. Les Editions Baghera, Paris (2019)
70. Von Goethe, J.W.: Die Metamorphose der Pflanzen. Friedrich Schillers Musen-Almanach. Hofbuchhändler Michaelis, Neustrelitz (1798)
71. Von Linné, C.: Philosophia botanica. R. Kiesewetter, Stockholm (1750)
72. Von Neumann, J.: The general and logical theory of automata. In: Cerebral Mechanisms in Behavior-The Hixon Symposium, pp. 1–41. John Wiley, New York (1951)
73. Waddington, C.H.: Morphogenesis and the field concept. Nature **138**, 809 (1936)
74. Waddington, C.H.: The Epigenetics of Birds. Cambridge University Press, London (1952)
75. Waddington, C.H.: Towards a Theoretical Biology, vol. 4. Edinburgh University Press, Edinburgh (1972)
76. Watson, J.D., Crick, F.H.: Molecular structure of nucleic acids; a structure for deoxyribose nucleic acid. Nature **171**, 737–738 (1953)
77. Williams, R.F.: Expanding attractors. Publ. Mathématiques IHES **43**, 169–203 (1974)

Jacques Demongeot is Professor Emeritus at the Université Grenoble Alpes (UGA) and Honorary Senior Member of the Institut Universitaire de France (IUF). He founded in 1979 the Team TIMB (Traitement de l'Information et Modélisation en Biomédecine) and in 1987 the CNRS Laboratory TIMC-IMAG (Techniques de l'Imagerie, de la Modélisation et de la Complexité) in Institut d'Informatique et de Mathématiques Ap-pliquées de Grenoble (IMAG). He directed TIMC-IMAG until 2011. In 1985, he founded the Medical Informatics Service at the Grenoble University Hospital. In 2003, this Service became a Public Health Center, which he directed until 2012. He founded in 1992 EDISCE (Doctoral School of Engineering for Health, Cognition and Environment), which he directed until 2003. From 1997 to 2000, he was responsible for Health Technologies at the Ministry of Research, then from 2002 to 2007 he directed the CNRS Complex Systems Program. He has published over 500 articles in applied mathematics, medical informatics and genetics. He is Member of the Chilean Academy of Sciences and Fellow of the International Academy of Medical and Biological Engineering (IAMBE). Email: Jacques.Demongeot@univ-grenoble-alpes.fr.

Chapter 11
Contributions of Morphodynamic Models in Medicine

Pierre Bounolleau

11.1 Introduction

Opening up a field as vast as medicine to theories stemming from fundamental mathematics is an unusual undertaking. Medicine today has a strong (and necessary) scientific dimension, but it remains above all a practice of care, an action-oriented discipline, a daily confrontation with the constraints of physiological complexity. The tradition of mathematical formalization is historically limited there.

Yet contemporary medicine, despite its intense theoretical and technical progress over the past two centuries and especially in recent decades, still has a number of theoretical blind spots. Having moved from the reductionist determinism of 19th century biology and physiology to the statistical thought of the second half of the 20th century, medicine no longer succeeds, for various long-term historical reasons that may relate to institutional culture and societal constraints, to progress in the management of certain clinical manifestations of physiological complexity due to a lack of conceptual framework and long-term theoretical vision. This concerns in particular the functional disorders, halfway between the traditional fields of competence of psychiatry and "organ medicine" ("somatic medecine"). These clinical situations share physiological matters that raise issues of disorganization of cellular and tissue networks which are particularly difficult to theorize. Their therapeutic management is difficult to implement, and is still limited today to empirical approaches with random results.

Catastrophe theory developed by René Thom from the 1960s onwards represented an important theoretical tipping point. It came from morphological thought, which is very old but has remained a minority in the history of science. Joining the vast body of theories of complexity, dynamical systems and connectionist cognitive sciences, which experienced considerable development at the same time, catastrophe theory

P. Bounolleau (✉)
302 Rue des Ormes, 45160 Olivet, France

© The Author(s), under exclusive license to Springer Nature Switzerland AG 2024
I. Marcos and C. Morier (eds.), *The Relevance of René Thom*, Lecture Notes in Morphogenesis, https://doi.org/10.1007/978-3-031-54982-3_11

paved the way for a morphodynamic program, proposing new types of scientific modeling based on the notion of equilibrium and structural stability.

This article proposes to open some lines of thought on the validity of this morphodynamic program for both theoretical and therapeutic research in the medical field.

11.2 The Tension Between Traditional Medicine and Science

Medicine is a discipline crossed, for as long as available historical data go back, by important methodological problems, linked to its dual status as a science both fundamental (theoretical) and applied (pragmatic). This is linked also to the extraordinary complexity of its object, the human organism. Medicine therefore lends itself particularly well to reflection on the notion of models in science.

One of the most constant and universal problems in medicine is, in particular, the intrinsic tension between two major types of methodological models, which can be called the *traditional* approach on the one hand, and the *scientific* approach on the other. This tension concerns both diagnostic reasoning and therapeutic practices. As far as Western civilization is concerned, this is closely linked to the profound methodological debates that science in general has been experiencing for as long as we can go back, around transversal concepts such as causality, emergence, form, and oppositions of the dualism/monism, reductionism/holism type. The description of these debates is particularly complex to order; it relates in particular to history and philosophy of science.[1]

The distinction between the two approaches can be understood as follows:

(1) the traditional approach schematically illustrates the *pragmatic* dimension of medicine, that is its constitutive mission of providing an effective solution to any clinical situation (disease, symptom, illness, disorder, trauma). This type of approach favors action, even if it has to sacrify at least part of the theorization; it tends to consider the organism and the person as a whole that is difficult to decompose. It thus refers to a holistic point of view, in which the relations between the parts have priority over the study of the parts themselves.

(2) the scientific approach, described above all within the framework of Western civilization but whose history is otherwise incompletely known to us, sets itself in contrast the task of modeling the human organism and its functioning as rigorously as possible, on the diagnostic and therapeutic levels. This approach may sometimes relegate to the background the practical effectiveness of care when the theoretical model fails. It is closely linked historically to biological reductionism,[2] attached to the study of *parts* thought of in a *unidirectional* way

[1] Andler et al. [1].

[2] Andler et al. [1], p. 946–947.

(according to a tree structure: organs or systems, subsystems, cells, cellular organelles, macro-molecules, molecules, atoms, etc.), more than to the relations between the parts.

The traditional therapeutic approach in medicine has been identified in just about every known human society and culture throughout history. There is no unified transhistorical or transcultural definition, but the following fundamental criteria can be retained: (1) a relational dimension, in the sense of an interaction between the patient and a therapist, (2) rich and varied sensory stimuli, organized in an elaborated aesthetics in terms of place, ritual, duration, mediating objects; (3) an empirical character, where the art of the therapist plays a central role, (4) an inscription in time.

11.2.1 Causalistic and Mechanistic Determinism in Medicine

The first historically known practitioners of science-based medicine, such as Hippocrates or Galen in Greco-Roman antiquity, set themselves the main task of introducing rigor and reproducibility into diagnosis and therapeutic choices.[3] The study of human anatomy and biology, the elimination of the religious dimension were among the main aspects of this intellectual impulse. Theoretical thinking continued to build up over the centuries, mainly around biology, particularly through progress in macroscopic and then microscopic anatomy. However, due to a lack of theoretical knowledge, therapeutic progress remained very limited until the 19th century.

The 19th century marked a major turning point in the history of medicine, with the rapid progress of physiology, which identified organs and their functions, and described diseases based on the dysfunctions of these organs. Advances in microscopy oriented the theoretical explanation towards reductionism, where the whole is deduced from the observable interaction of the parts. Under the leadership of physics and classical mechanics of previous centuries (Descartes, Newton), and despite the powerful tensions that divided scholars,[4] the notion of simplified, "mechanistic" causality became central to medicine at this time, supported by the great figures of Western physiology like Claude Bernard. Treating a patient was a question of identifying the unique or main "cause" of the disease, and of acting on this cause. This type of methodology was mainly developed on the model of infectious diseases, whose causal sequence was perceived as easy to demonstrate empirically and to treat.

This deterministic, causalistic and mechanistic methodological framework occupied the forefront of scientific reflexion in medicine throughout the first half of the 19th century. Its consequences are of several orders:

[3] Falissard [3].

[4] Andler et al. [1], p. 825–938 "La causalité".

(1) although it has enabled major advances in the understanding of physical, cellular and biochemical processes, it has come up against the insurmountable difficulties of modeling for complex phenomena such as cognition and language. In fact the physical matter of such complex phenomena is neither sensorially observable nor empirically accessible.[5] Their study has therefore been constituted according to logico-formal and structural approaches.[6]

(2) modern medicine, like other disciplines, has been very strongly marked by these theoretical constraints from its origins. The institutionalization of its care sectors in the 19th century was done by organ, in the model of reductionist physiology by hierarchy of levels. The dualistic gap that was already present since the 16th century progressively widened between medicine of the body ("somatic" medicine) associated with causalist biology, and medicine of the mind ("psychic" medicine) associated to formal idealism. Such a split has also taken place in other university fields, where meaning and language separated from what is known as "hard" sciences of physico-chemical inspiration and joined the new "human" sciences (philosophy, linguistics, sociology).[7] Similarly, pathologies of cognition and behavior (apart from the restricted case of aphasiology) drew little or no benefit from advances in organ medicine, and were managed by care disciplines that follow the same logico-formal path, like psychiatry, clinical psychology, psychoanalysis. This led, among other things, to the separation of neuropsychiatry in the 1960s into psychiatry (disorders relating to the "mind") and neurology (disorders relating to the "body").

(3) despite the conceptual evolutions that appeared more recently, in the second half of the 20th century, causalist and reductionist determinism still deeply permeates the thought of the human body and care, both in the general population and among practitioners of all professions. The model of molecular biology, which is the recent form of this determinism, is still omnipresent in daily medical practice, in physiological reflection as well as in therapeutic processes, often at the expense of approaches relating to complex biological equilibrium.

11.2.2 The Move Towards Statistical Thinking

The causalist reductionism of the 19th century gradually proved to be unsuited to a large number of daily medical situations, and in particular to chronic diseases. It appeared that beyond infectiology, understanding the origin of a disease, although useful both theoretically and practically, was not always possible with a sufficient degree of precision, and often did not allow effective treatments to be implemented. Another methodological approach became necessary.

---

[5] The pioneer works of neuro-histology in the XIXth century. (Ramon y Cajal, Golgi) discovered the existence of neurons, but without saying anything about their dynamic functionning.

[6] Petitot [12], p. 16 "matière, substance, forme: la naturalisation du structuralisme".

[7] Petitot [10], p. 97; these different logico-formal paths evolve in the XXth century towards the structuralist path, to which Lacanian psychoanalysis joins.

Statistical models, resulting in particular from the work of mathematicians such as Fischer, Neyman and Pearson in the first half of the 20th century in agronomy (seed selection), became widely accepted in the 1950 and 1960s to become the methodological standard of scientific medicine. The rigorous comparison of groups of randomized patients made it possible to circumvent the complexity of the biological mechanisms underlying a disease and a treatment, in order to evaluate the effect of therapeutics in a pragmatic but nevertheless rigorous and standardized way.[8]

For several decades now, clinical trials have thus largely entered the daily lives of doctors and patients alike. This represents a decisive methodological break with the general intellectual framework of 19th century medicine, which gave precedence to theoretical modeling in the strong sense, that is to say a conceptualization and specific logical sequences of biological and physiological objects. The choice of statistics, on the other hand, renounces at least in part this specific modeling deemed unfeasible because of the excessive complexity of the field of study, and turns for pragmatic reasons towards a mathematical formalization of a general type, universal in scope, independent of human biology. Of course, all modeling in the strong sense is not abandoned: therapeutic research works largely from the corpus of data coming from fundamental biological research, and these different lines of research articulated between them practice a work of theoretical modeling in the sense of *reconstructing* phenomena specific to their field of study. However, the final realization of decision-making processes in the field of therapy relies, through statistics, on a formalization unrelated to the biological object, which is no longer strictly speaking a modeling.[9]

11.2.3 The Dichotomy of Care: Scientific/Traditional

This aspect of intellectual and scientific history has important implications in collective culture. It implicitly but profoundly permeates the actors of medicine, digging a conceptual divide between scientifically validated care and traditional care.

Statistical methodology, which entirely structures the modern conception of intellectual rigor in medicine, has in fact been very largely constructed by and for easily reproducible, industrializable care, namely drugs. All other types of care, which are sometimes referred to as "artisanal" care, are much more difficult to assess by statistics and controlled studies. Apart from the case of surgery, which involves both traditional care (with inter-operator variability) and industrial care (with specific equipment, rigorous procedures), "artisanal" (traditional) care is essentially represented by *relational therapies*, that is to say all type of care that primarily involves

[8] Given the duration and cost of the optimal clinical trial design (prospective, controlled, randomized, double-blind), other methods of treatment evaluation may be used, with varying levels of evidence: prospective open-label trials, retrospective trials, cohort studies, guidelines from authorities and/or expert committees.

[9] Petitot [11], p. 75–76, on the question of mathematical modeling in the sciences in general.

the singular dialogue between a patient and a practitioner (which in clinical trials is called *placebo*), without the use of institutionally validated medication.[10]

Medicine is therefore constantly oscillating between several contradictory needs, that of taking care of a "real" patient as a whole, but one that cannot be modeled because of its complexity, a "biological" patient who is the imperfect but nevertheless rigorous theoretical model created by science (physiology and human biology), and a "statistical" patient resulting from clinical trials and defined by mean, median or standard deviation type tools, to circumvent the difficulties of theoretical modeling.

For medicine, the challenge is to guarantee its own reliability and maintain trust between patients and practitioners, by offering theorized and rigorously evaluated care, while at the same time dealing with the biological complexity of the body, which sometimes thwarts this care. The room for maneuver is narrow, with the risk of rejecting in the field of obscurantism or mystery all that part of the care which is nevertheless indispensable and which concerns the human relationship.

These debates have been partially transformed since the beginning of the 21st century, with the appearance of new diagnostic terms and new care networks: conceptual categories of the lesional/functional type, better adapted to biological complexity, are gradually (but slowly) replacing the classic somatic/psychological dichotomy, which is often poorly experienced by patients; transversal care networks (pain assessment and treatment centers) are being developed to take over from the traditional organ specialty channels in the management of non-injury (non-"somatic") disorders. Institutional and university health actors are taking up the scientific (in fact essentially statistical) evaluation of *alternative* or *complementary* therapies.

At present, however, neither scientific medicine nor research in human biology and physiology yet has reliable theoretical models specific to their objects, in order to conceptualize the diagnostic and therapeutic means of traditional medicine. There persists a lack of major scientific models on the laws of operation of the "meso" ("mesoscopic") level of cellular networks, which is the intermediate anatomical-physiological level between the two levels historically most accessible to scientific study—the microscopic level of cellular and molecular biology and the macroscopic level of the organ.

11.3 Catastrophe Theory and Morphodynamic Models

Morphodynamics is a vast set of theories and methodologies that can be to approached through what is undoubtedly one of the most successful attempts at formalization, René Thom's catastrophe theory. This theory is based on highly technical mathematical foundations, which make it particularly difficult to access for non-mathematicians, and even, to a certain extent, for mathematicians unfamiliar with differential geometry. Nevertheless, by the audacity of its intuitions, the fruitfulness

[10] Falissard [4].

of its hypotheses and the variety of its objects, it has achieved a turning point in the recent history of scientific modeling.

René Thom has worked from the very beginning of his career on issues relating mainly to geometry and topology. He joined the National Center for Scientific Research[11] in Strasbourg in 1946 at the end of his studies, and began to work on algebraic topology, at a time when this discipline was experiencing very significant technical development. He became interested in the notion of boundary (french *bord*) and developped the theory of cobordism, which will be rewarded with the Fields medal in 1958.

At the beginning of the 1960s, starting from an experiment on the variations of envelopes of light rays (caustics) in optics, René Thom sought to go back to the mathematical laws (singularities) underlying the appearance of boundaries, also called phenomenological discontinuities, delimiting the natural forms.[12] He gradually built what he called a "semi-philosophy", which he extended to various fields of knowledge, about how a material substrate, whatever its level in the hierarchy of the non-living or the living (atom, molecule, cell, organism, up to the society of individuals), self-organizes spatially and temporally, under the effect of external and internal constraints, according to a limited number of paths and forms among the vast range of possibilities. The term phenomenological discontinuity will then be replaced by that of catastrophe, giving rise to *catastrophe theory*.

It constitutes a theory of observable (phenomenal) form, which it seeks to articulate with the dynamics of "accidents" (discontinuities) of the underlying matter, by going back to the mathematical laws of structure (singularities); the basic conceptual chain is therefore: form => structure => discontinuity => singularity.

Catastrophe theory renewed and enriched in depth the notion of space (object of topology) and continuity; through its transversal and qualitative dimension, it overhauled the historical distribution of certain theoretical tools between different disciplines, for example by opening up the theory of singularities, hitherto known only from physics, to distant fields of study. In this way, it questioned the intellectual traditions of many fields of knowledge: reductionism in the case of biology, logical and structural formalism in the case of the human sciences.[13]

Catastrophe theory fits fully into the long tradition of theories of *form* and *structure*: *form* arises from the dynamic evolutions of matter to constitute what we perceive empirically (sensorily, cognitively); the *structure* is the abstract theoretical model (the internal and external relations of the form) that one reconstructs throughout the process of knowledge.

These theories of form recognize Aristotle's philosophy as the main historical starting point and fundamental foundation. They were profoundly rethought by Leibniz at the turn of the 17 and 18th centuries, then by Goethe and Kant; they were extended and developed in the aera of life by several great biologists of the 19 and 20th centuries (Geoffroy Saint-Hilaire, Waddington, D'Arcy Thompson), then

[11] French Centre National de la Recherche Scientifique (CNRS).

[12] Thom [15], § "La Théorie des catastrophes: genèse", p. 20–29.

[13] Thom [14], p. 83 and 86.

in the disciplines of language and in anthropology by the great figures of scientific structuralism (C. Lévi-Strauss, R. Jakobson, L. Hjelmslev, A.J. Greimas).

Catastrophe theory is also part of the major scientific shift in theories of complexity and self-organization of the period 1960–1980, during which several research groups working in various disciplinary fields[14] converged towards the same idea of using the theories of singularities, bifurcations and form structuring processes in general.

René Thom's proposals and work paved the way for a research program around the key concept of *morphodynamics*, whose intention is to achieve a naturalist theory (a mathematical modeling) of the processes generating natural forms.[15] This program deeply renews the form/matter dualism of Aristotle's thought, by developing the founding idea that cognitively perceived forms do not arise from sets of autonomous relations (i.e. *structures*) grafted on an amorphous matter that would be ontologically foreign to them, but on the contrary do they emerge to our perception through processes of dynamic self-organization of matter.[16]

The morphodynamic approach therefore influenced the philosophical (epistemological) foundation of disciplines dealing with individual and collective cognitive processes, notably the historically "structural" disciplines such as linguistics, narratology, literary studies, semiotics, which were originally built upon formalist and logico-algebraic bases of a dualistic (neo-Aristotelian) nature. This renewal affected by extension all the theories of the first "cognitivism", built in the years 1960–1970 on the dualist bases of the computer science approach (*software/hardware*), and participated in the passage towards second-generation cognitive sciences, known as "connectionists".

There is today, in the 21st century, a massive convergence of morphodynamic models and cognitive sciences around the program of naturalization of mental processes (language, behavior, reasoning, learning), aimed at studying how these processes emerge from self-organized dynamics within neural networks.

11.4 Applications of Morphodynamics in Medicine

11.4.1 Basic and Clinical Research: A Few Examples

Theoretical cognitive neurosciences have largely integrated since the 1980s the connectionist and morphodynamic concepts of cellular basins of attraction in the study of the functioning of mental processes. Within a cognitive (artificial or biological) system, a *basin of attraction* (distributed around an *attractor*, which is its energy minima and organizing center) is a stable configuration of an neuronal

---

[14] Under the aegis of people like René Thom and Henri Atlan in France, Ilya Prigogine in Brussels, Hermann Haken in Germany, Christopher Zeeman in England.

[15] Petitot [10], p. 1.

[16] Petitot [10], p. 2 and 3.

assembly,[17] which can be associated with the biological support of an idea (a word, an abstract perception, a representation), whatever its level of extension. These basins of attraction have, in particular, a very important function of energy saving, the cells composing them being at metabolic "rest". They also have specific properties of dynamic evolution, for example oscillatory capacities.

The penetration of morphodynamic models is an important shift in the history of these disciplines, insofar as they bring for the first time effective mathematical solutions to hitherto insurmountable technical obstacles in the theorizing of the functioning of the intermediate ("mesoscopic") anatomical and physiological levels between the cellular level (accessible to microscopic observation) and the macroscopic level of the organ, mentioned in the first part of this article.[18]

It is also in research on protein folding and conformation that the influences of morphodynamic models are the most sensitive. Hypotheses on the presence of certain macro-molecular forms rather than others, on the self-organization of these protein macro-molecules in space and on their dynamic interaction with the biological environment, have already incorporated models based on potential functions and basins of attraction since the 1990s.[19] Key concepts of catastrophe theory such as structural stability, spatial dynamics of forms and phase transitions are strongly present.

This field of study is not strictly limited to basic research, and includes important clinical applications. In particular, morphodynamic protein models enabled decisive progress in the understanding of prion-associated diseases in the 1990s, and continuously since then, in the study of neurodegenerative diseases (Parkinson's syndromes, Alzheimer's type cortical degenerations). This concerns in particular the diffusion and the dynamic equilibirum within the nervous system of the malformed protein inclusions associated with these diseases.

11.4.2 From Semio-Linguistic Morphodynamics to Traditional Medicine

At the end of the 1960s, René Thom developed applications of catastrophe theory in linguistics and semiology, introducing a hypothesis of *concept geometry*, where the

[17] Dehaene [2], p. 244.

[18] Petitot [12], p. 16 (*we translate*): "As far as natural morphologies are concerned, there are now physico-mathematical models. Theories [...] explain how, on the basis of interaction processes and coordinated collective behaviors (cooperation and conflicts) located at an intermediate („mesoscopic") scale, small-scale („microscopic") units can organize themselves into emergent large-scale („macroscopic") structures. These theories [...] have shown that the emerging „macro" morphological level is essentially organized around the *singularities—qualitative discontinuties*—of the underlying „micro" processes".

[19] Holm et al. [5].

syntax corresponds to the set of relations between the attractors (minima of potentials), and semantics to the morphology of basins of attraction.[20] Using mathematical tools (archetypal morphologies, actancial graphs), he established a close link between cerebral structures of perception that are phylogenetically very old, and the basic grammatical structures common to all known languages (verbal valences, cases of declension, actancial roles of the sentence). Converging with the results of psychology and cognitive linguistics, he joined the localist hypothesis of language on the anteriority of the concept of space in syntactic and semantic structures, and on the anchoring of language capacities in very primitive perceptual structures of the human brain.[21]

René Thom also conducted research on the extension of the localist theory of language to narrative theories, and on the fundamental anchoring of narrative dynamics in perception. This research was continued by his successors, notably Jean Petitot, who extended its scope in a morphodynamic rereading of structuralism, according to which "integrated cognitive neurosciences hold the key to the scientific future of structuralism".[22] This rereading opens, among other things, important lines of research for the theorization of traditional medicine practices.

For instance, C. Lévi-Strauss in anthropology, A-J. Greimas in linguistics and semiotics, and C. Morazé in history of civilizations, who are part of this intellectual legacy of morphodynamic structuralism, have thoroughly studied a large number of narrative, mythical, religious, ritual, gnostic or divinatory practices in different civilizations throughout the centuries. These practices have in common to bring into play in a constitutive way apparently simple logico-combinatorial structures (numerical, algebraic and/or geometric) to "cerebrally" manage (i.e. simplify) the immense complexity of human life.[23]

The implementation of these logico-combinatorial structures consists in associating with their basic formal elements data from lived experience. These data can be very general (existence, life, death, position of the stars) when it comes to a great collective narrative like a myth, or more local, related to the life of an individual (his network of social or environmental relations). These external data will be amplified and enter into resonance within the "grammar" of the structure. This resonance automatically produces relationships with other data from the life context initially used, inducing a sort of "stabilization" of this context, that is a generation of *meaning*. In traditional divinatory practices, for example, these amplified, automatically produced relationships ("meanings"), are used to interpret the local situation or context that provided the elements in the first place.

C. Lévi-Strauss provided in a 1949 article, *L'Efficacité Symbolique* (The Symbolic Efficiency), a remarkable example of a similar logico-combinatorial theorization for the functioning of a traditional medical practice. He describes the intervention of a shaman who, called to the bedside of a woman in difficult childbirth, uses a

---

[20] Petitot [9], p. 26–27.

[21] Petitot [9], p. 33–35.

[22] Petitot [12].

[23] Petitot [8].

story reproducing by role-play the organs and the topographical configuration of the physiological event in progress. The structured and richly decorated aspect of his intervention (narrative, singing, postures, gestures, facial expressions, intonations, costume, accessories...) allows to:

> going from the most banal reality to myth, from the physical universe to the physiological universe, from the external world to the internal body. [...] The technique of storytelling therefore aims at restoring a real experience, where the myth merely substitutes the protagonists. [...] The cure would therefore consist in making thinkable a situation first given in affective terms, and mentally acceptable, pains that the body refuses to tolerate. The fact that the mythology of the shaman does not correspond to an objective reality is of no importance: the patient believes in it, and she is a member of a society which believes in it. [...] What she does not accept are incoherent and arbitrary pains, which constitute a foreign element to her system, but which, by appealing to the myth, the shaman will replace in a whole where everything fits together. [..] This is the passage to this verbal expression (which allows, at the same time, to live in an orderly and intelligible form a current experience, but, without that, anarchic and ineffable) which causes the unblocking of the physiological process[24]

C. Lévi-Strauss formulates the hypothesis that the narrative (and the associated material dimensions), used as a mediator of the therapeutic relationship, has the capacity to produce relationships between events where they are lacking, or to amplify insufficient relationships, and to act in the patient's brain by rearrangement towards stability of hitherto disordered elements of his life context. The story manages to deform and manipulate the reality it reproduces to infinity, and this grammar responds to a delicate balance whose handling is the whole art of the therapist. The therapist then has a role of creator in connection with the cultural codes of the society in which he lives.

It should be noted that the morphodynamic approach of structuralism has extended its rereading of classical texts to many other fields (painting and plastic arts, literature, poetry, novels, tales), exploring the different spatial and temporal dimensions of sensory perception, in order to work out what could be called a global theory of *aesthetic efficiency*, in the sense of the intrinsic stabilizing properties of the forms of perception.[25]

Morphological models of complexity, and in particular the work on the resonance of dynamic networks of neural oscillators, bring here a decisive advance: it becomes possible to "naturalize" the data from anthropology, history, literary and visual arts, in other words to confirm their theoretical intuitions and anchor them in "neural matter".

At the same time, work in recent decades[26] has revealed very specific properties of consciousness as a brain function. Consciousness represents the top of the functional tree of the nervous system, and the central node of decision-making processes. It is capable of providing the brain with an illusion of perceptual unity, of coherence of the flow of thought, which has a profoundly adaptive value; it allows the body to automatically fill in the gaps in the perception of its environment, to optimize its

[24] Lévi-Strauss [6], p. 13–19 (*we translate*).

[25] Petitot [13]

[26] Naccache [7]; Dehaene [2].

energy regulation and therefore its decision-making capacities. The "price to pay" for this perceptual synthesis is the permanent and dynamic production of a very powerful internal fiction.

This fictional capacity seems phylogenetically very old. It has been identified and theorized in physiological condition in recent decades, but it has been known and well documented in pathology since at least the 19th century. There are indeed frequent clinical situations where the injured brain is no longer able to analyze reality in a coherent way, and replaces it in whole or in part with self-generated fictional scenarios that are sometimes very elaborate: delusional states of acute or chronic psychosis; elective lesions of the internal temporal cerebral regions involved in memory storage, of degenerative origin (Alzheimer's disease) or toxic origin by chronic alcoholism (Korsakoff's syndrome).

Based on this clinical research data, cognitive science work has focused on the use of these fundamental fictional properties of the brain in certain relational therapeutic practices such as psychoanalysis:

> By inventing the unconscious, Freud brought to light an essential cog in our consciousness: precisely this vital need to interpret, to give meaning, to invent through imaginary constructions. Today we are beginning to know the cerebral reality of these mental fictions, which govern our conscious thought. [...] Freudian psychoanalysis seems to me to convey this art of composing our existence in the form of this constantly revised novel that we never finish writing[27]

These data can be brought closer to the comparative analysis between shamanism and psychoanalysis carried out by C. Lévi-Strauss in *L'Efficacité Symbolique*: both methods

> aim to bring about an experience; and both achieve this by reconstituting a myth that the patient must live, or relive [...]. Each time it would be a question of inducing an organic transformation, consisting essentially of a structural reorganization, by bringing the patient to live intensely a myth, sometimes received, sometimes produced, and whose structure would be, at the level of the unconscious psyche, analogous to that whose formation one would like to determine at the level of the body. Symbolic efficiency would consist precisely in this "inductive property" possessed, in relation to each other, by formally homologous structures that could be built, with different materials, on the different levels of the living[28]

By synthesizing these different mathematical, semio-narrative, anthropological and neuro-cognitive sources, it seems possible to formulate a hypothesis on the intrinsic fictional properties of the brain intuitively used by traditional medical practices:

- the human brain is a complex system radically out of equilibrium, a condition of both its prodigious adaptability and its inherent fragility; this physiological property imposes very powerful compensatory stabilization mechanisms;
- narrative creativity, a phenomenon universally observed in all known human societies and at all levels of collective and individual production, and confirmed

[27] Naccache [7], p. 439; (*we translate*).

[28] Lévi-Strauss [6], p. 21 and 24; (*we translate*).

in pathology by medical experience, is one of these cerebral compensatory mechanisms;

- Therapeutic practices such as psychoanalysis, physio-anatomical geometries (acupuncture), artistic therapies and many others have long intuitively and empirically used these dynamic properties of neural networks and in particular the ability of the brain to generate coherent narratives and fictions; these therapeutic practices are based for these purposes on elaborate aesthetics, whose elementary components are devoid of their own material value, but possess by their positional value and their internal interplay a capacity for resonance, that is able to reorganize the patient's experience and the perception he has of his body in a stable and coherent state or history;
- catastrophe theory and its semio-linguistic applications provide mathematical models consistent with this hypothesis; in particular, actancial graphs and *archetypal morphologies*, as very deep fundamental structures of human cognition well documented experimentally, are involved in the functions of perception, language, as well as action and behavior.

11.5 Conclusion

We have looked at how, because of the specificity of its objects and its particular history, the practice of medicine is still often caught up in a body-mind dualism ill-suited to its cultural evolutions, and in a conflicting dichotomy between the reductionism of molecular biology, unified and institutionally powerful, and the empirical holism of traditional therapies, fragmented among innumerable practices. Statistical models, despite the undeniable therapeutic progress they have enabled, do not provide conceptual insight into the substance. The experience of care among the population as well as among caregivers continues to oscillate between the demand for scientific rigor and the demand for intuitive efficiency specific to the complex relational dimension of care.

We have suggested, through a few lines of thought, how morphodynamic models have advanced medical sciences, around the notions of dynamic equilibrium of cellular networks, structural stability and spaces of forms organized as attractors. Such concepts are slowly spreading in the institutional culture, without yet becoming really perceptible in the organization of daily care.

We believe that through them there is the possibility of developing rigorous models, not limited to simple statistical evaluation, of the relational dimension of care and of the aesthetic techniques (in the sense of organized perceptive sets) on which it is based, and consequently of bringing together the scientific part and the traditional part of medicine.

René Thom worked on mathematizing the way in which nature creates discontinuity (and hence forms) from a continuous background. Catastrophe theory and the morphodynamic program that surrounds and extends it, without seeking to resolve all the oppositions, propose a reflection on the dualisms of different natures that often limit the scientific approach, and new ways to overcome them. By theorizing the continuum, they put transdisciplinarity at the heart of their ambitions.

References

1. Andler, D., Fagot-Largeault, A., Saint-Sernin, B.: Philosophie des sciences I et II. Gallimard, Paris (2002)
2. Dehaene, S.: Le Code la Conscience. Odile Jacob, Paris (2014)
3. Falissard, B.: Thérapies complémentaires: cadre méthodologique et éthique. In: Formation spécialisée transversale pharmacologie, thérapeutique et douleur. Online conference (2019)
4. Falissard, B.: Les méthodes de la recherche clinique exigent-elles en psychiatrie des adaptations spécifiques? L'Information psychiatrique **85**, 153–159 (2009)
5. Holm, L., Sander, C.: Mapping the protein universe. Science, **273**(5275): 595–603 (1996)
6. Lévi-Strauss, C.: L'efficacité symbolique. Revue de l'Histoire des Religions, tome **135**(1), 5–27 (1949)
7. Naccache, L.: Le Nouvel Inconscient. Freud, le Christophe Colomb des Neurosciences. Odile Jacob Poches, Paris (2009)
8. Petitot, J.: La notion de structure chez Charles Morazé. Oral intervention, Maison des sciences de l'homme, Paris (2005)
9. Petitot, J.: Les Premiers textes de René Thom sur la morphogenèse et la linguistiques, pp. 1966–1970 (2015)
10. Petitot, J.: Syntaxe topologique et Grammaire cognitive. Langages **25**(103), 97–128 (1991)
11. Petitot, J.: Nature et enjeux de la modélisation en sciences sociales. Le Genre humain **1**(33), 75–102 (1998)
12. Petitot, J.: La généalogie morphologique du structuralisme, pp. 97–122. Critique, Paris (1999)
13. Petitot, J.: Morphologie et Esthétique. Maisonneuve et Larose, Paris (2004)
14. Thom, R.: Paraboles et catastrophes, Flammarion, rééd. Coll. Champs, Paris (1983–1989)
15. Thom, R.: Prédire n'est pas expliquer. Flammarion, Paris (1993)

Chapter 12
Some Remarks on Catastrophe Theory and Biology

Sara Franceschelli

12.1 Introduction

In the founding article of catastrophe theory, Thom states that the sources of his theory are twofold: on the one hand his research in topology and differential analysis, on the other hand the treatises of embryology, in particular the books of Conrad Hal Waddington and his notions of "epigenetic landscape" and "chreod". These two notions were introduced by Waddington through images, even before definitions. In this chapter I defend the idea that the relation of catastrophe theory to biology, which—by Thom's admission—cannot be subjected to experimental control, rests on the bet of its relevance as an expression in mathematical terms of the images constituting Waddington's epigenetic landscape.

After showing in what sense the notion of structural stability allows Thom to relate his theory to the epigenetic landscape, I consider the limits of this ambition but also the perspectives opened by the Thom-Waddington dialogue in the context of the recently renewed interest in Waddington's voice as a theorist of developmental canalization. In the first section I will introduce the notion of structural stability in the framework of Thom's structuralism. In the second section I'll discuss the system of differential equations through which Thom's introduces catastrophe theory, written in reference to the Delbrück's model of cellular differentiation, and that take initially the form of reaction–diffusion equations. In the third section I'll consider the dialogue between Thom and Waddington with respect the possibility of grasping through equations the dynamics properties of equilibria of the epigenetic landscape. In the fourth section I consider the criticisms of Waddington concerning both Turing's and Thom's approaches to the mathematisation of the property of development focusing on the property of canalization, considered by Waddington as an essential trait of organisms.

S. Franceschelli (✉)
ENS de Lyon, IHRIM and IXXI, Lyon, France
e-mail: sara.franceschelli@ens-lyon.fr

© The Author(s), under exclusive license to Springer Nature Switzerland AG 2024

I. Marcos and C. Morier (eds.), *The Relevance of René Thom*, Lecture Notes in Morphogenesis, https://doi.org/10.1007/978-3-031-54982-3_12

12.2 Structuralism and Structural Stability

For Thom, there is an analogy between topology as a mathematical theory, whose main problem is to construct a global form from local properties, and theoretical biology, which, from a local biochemical determinism, seeks to explain the stability and reproduction of global spatio-temporal structures. Catastrophe theory allows us to think about the transition between local properties and global forms in a non-reductionist way, in the sense that it does not attempt, as in the reductionist approach, to reconstruct a complex spatial form from simple elements deemed to belong to a more fundamental theory. Thom's catastrophic morphology, on the other hand, is structuralist—and this is precisely where its non-reductionism lies. On the structuralist enterprise Thom writes, in the collection of articles *Mathematical Models of Morphogenesis* published in 1980:

> Elle [l'entreprise structuraliste] représente sur le plan qualitatif l'homologie du célèbre "Hypothèses non fingo" de Newton. On se propose en effet, dans l'optique structuraliste, non pas à expliquer une morphologie à l'aide d'éléments empruntés à une autre théorie supposée plus fondamentale, ou plus élémentaire (comme d'expliquer la Biologie par la Physico-Chimie, ou la Sociologie par la Psychologie ou la Biologie), mais simplement d'améliorer la description formelle de la morphologie empirique, en mettant en évidence ses régularités, ses symétries cachées, en exhibant son caractère global par la description d'un processus formel qui l'engendre axiomatiquement. En ce sens, le structuralisme est une théorie modeste, puisque son seul but est d'améliorer la description ([16], 142–43)

Within this structuralist framework, the relationship of catastrophe theory to experience, and in particular to embryology, is not direct, but is achieved through the notions, introduced by the embryologist Conrad Hal Waddington, of "epigenetic landscape" and "chreod".[1] As we are going to see in the Sect. 12.3, these two notions were introduced by Conrad Hal Waddington through images, even before definitions. In this chapter I defend the idea that these images offer a privileged perspective for understanding the type of structuralist intelligibility of embryonic development proposed by Thom. It is no exaggeration to say that the relationship of catastrophe theory to biology as it emerges from Thom's early texts is crucially linked to its relevance to the mathematisation of these images, through the notion of structural stability.

Before considering these images in the Sect. 12.3, I wish to focus here on the shift from the notion of determinism to that of structural stability that characterises Thom's approach to morphogenesis. Thom notices that phenomena that cannot be considered to be determined in the Laplacian sense (as thus for Thom are to be considered phenomena of embryological development) are nevertheless mainly predictable if one has a knowledge of embryological treatises on development. It is thanks to considerations of structural stability that Thom can theoretically account for this predictability. This property of structural stability is, moreover, independent of the knowledge of the substrate and the forces at play, in full agreement with his criticism

[1] By the introduction of this neologism from the Greek Waddington intends a "necessary path of development".

of a reductionist approach to morphogenesis. Structural stability is the conceptual and mathematical tool that allows him to address the question of stability and change in living structures and thus to establish a link with embryology (as we shall see, via the epigenetic landscape).

Thom takes the notion of structural stability from the theory of dynamical systems, where it was introduced in physics and engineering to study self-oscillating systems.

For Thom a process (P) is structurally stable, if a small variation of the initial conditions leads to a process (P') isomorphic to (P") ([18], reprinted in Thom [16], 12). From a mathematical point of view the qualitative idea is that a function *F is structurally stable* if, by perturbing the function F sufficiently little, the perturbed function $G = F + \partial F$ still has the same topological form as the initial function F.

The notion of structural stability used by Thom was first proposed by Andronov and Pontrjagin in the 1930s [2] and is used to analyse oscillating physical systems in the 1937 *Control Theory and Self-Oscillations* by Andronov et al. [3]. Structural stability is described as a property that a dynamic system (in the sense of a mathematical model) should have in order to correspond to an observable system in the physical world: when writing an equation, one cannot take into account all the factors that influence a physical system and, moreover, it is not guaranteed that the factors considered remain perfectly constant during the evolution of the system. The notion of structural stability has an epistemic ambition: that a mathematical theory as such should be able to guarantee the quality of a model, i.e., its ability to correspond well to a given phenomenon.

The following passage is taken from the book co-authored by A. A. Andronov. In this text the expressions "rough systems" or "coarse systems" are used to designate what we now call "structurally stable" systems, following the first English translation supervised by Solomon Lefschez:

> What properties must dynamic systems (models) possess to correspond to physical systems?
>
> In setting out the differential equations we cannot take account of all the factors that influence in some manner or other the behavior of the physical system. On the other hand, none of the factors taken into account can remain absolutely constant during a motion of the system [...]. A number of parameters corresponding to physical parameters of the problem occur in the functions P and Q of our system equations, so these functions are never known exactly. Small variations of these parameters must leave unchanged the qualitative structure of the phase portrait. Therefore, if certain qualitative features appear for well-determined quantitative relations between the parameters but vanish for an arbitrarily small variation of the parameters, then it is clear that such qualitative features are not, generally speaking, observed in real systems. It is natural, therefore, to separate the class of dynamic systems whose topological structure of the phase paths does not vary for small variations of the differential equations. We call such systems "*coarse*" or *structurally stable* [...] (Andronov and Pontrjagin [2], Andronov et al. [3], 374–375).

This methodological perspective has been consolidated in the early 1960's by a global analysis theorem proved by a student of S. Lefschetz, Mauricio [12]. However, it has been shown by Stephen Smale that Peixoto theorem does not generalize for dimensions higher than two.[2] Taking into account these limitations to the ambitions of the program of a general classification of the dynamical systems on the basis of their properties of structural stability, Thom restricts the class of systems likely to be concerned by the theory of elementary catastrophes to gradient systems, but keeps intact the enormous epistemic ambition behind the notion of structural stability.[3] He even exacerbates it, explicitly invoking a Platonic-type argument, which seems to attribute to mathematical theorising alone, and prior to or even independently of any contact with experience, the task of discriminating good from bad mathematical models. According to Thom, a morphogenetic field on an open U of space–time resides in the data of a "universal model" of which the particular process under study is a copy. Such a process would unfold in accordance with the universal model given a priori and would, therefore, be structurally stable.

For Thom structural stability remains a property of any observable form, despite the vicissitudes of its mathematical definition. It is through structural stability, which should make possible to mathematically translate the properties of the epigenetic landscape, that Thom tackles the problem of embryonic development. Before considering, in Sect. 12.3, the notions of epigenetic landscape and chreod, and their relation with catastrophe theory, we are going, in next Sect. 12.2, to consider the system of differential equations wrote by Thom to give an illustration of catastrophe theory in biology.

12.3 The First Mathematisation: Reaction–Diffusion Equations for Delbrück's Model

In order to illustrate qualitatively an application of catastrophe theory, Thom explicitly writes a system of differential equations from Delbrück [5] model of cell differentiation, which Delbrück had introduced by a diagram without accompanying it with any equations as a commentary during a conference on genetics held in Paris in 1949 [5].

During the Paris conference a proof was presented for the heritability of certain phenotypic traits in paramecia and other ciliates over a large, but finite, number of generations.

This could be taken as evidence against the chromosomal transmission of these traits. G.H. Beale then advanced the hypothesis of the existence of populations of cytoplasmic genes, or plasmagenes, that would be transmitted through a finite number of cell divisions before disappearing, and which would be responsible for the existence of these traits. Delbrück wanted to show that the same observations could be

[2] For a discussion on structural stability see Smale [14]. The reader interested in the epistemological questions opened by the mathematical study of structural stability opened by Thom's program in the 60's should refer to Chaperon [4].

[3] On this question, we can refer to Thom [17].

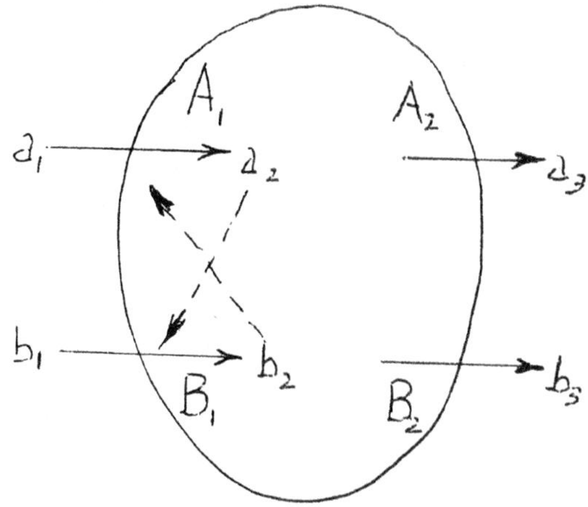

Fig. 12.1 Delbrück's model. A1, A2, B1, B2 are different types of enzymes inside the cell (which is represented by the circle). a1 and b1 are substances in the environment. a2 and b2 are intermediate metabolites produced from a1 and b1 under the influence of A1 and B1. They are in turn the substrates of enzymes A2 and B2, which transform them into a3 and b3, waste products. Entirely redrawn on the basis of Delbrück [5]

explained without resorting to genes or plasmagenes. To this end, he put forward the following argument ([5], p. 33, our translation):

> ... many systems in flow equilibrium are capable of several different equilibria under identical conditions. They can move from one equilibrium state to another under the influence of transient perturbations

Delbrück thus proposed a model of interacting metabolic pathways, as illustrated in Fig. 12.1, but without explicitly writing the associated equations.

A1, A2, B1, B2 represent different types of enzymes within the cell (represented by the circle). a1 and b1 are substances in the environment, while a2 and b2 are intermediate metabolites produced from a1 and b1 under the influence of A1 and B1. They are, in turn, the substrates of enzymes A2 and B2, which transform them into a3 and b3, which are waste products. In a constant environment, this model always remains in a stable state. But at this point, Delbrück added the hypothesis that there exist mutual interactions between the two series of enzymatic reactions (shown in the diagram as dotted arrows). His diagram evokes cross-reacting metabolic pathways. Now there exist three possible equilibria for the same conditions of the environment: two stable states and one unstable.

Commenting on this model and the role it has played in the development of developmental biology, Evelyn Fox-Keller observed that it has undergone a series of semantic shifts according to the contexts in which it has been considered, acting more as a metaphor than as a model [7].

In order to mathematically grasp the processes described by Delbrück's model of cellular differentiation, Thom, introducing k chemical substances s_1, s_2, ... s_k of respective concentrations c_1, c_2, ... c_k, writes the following differential law of variation of concentrations:

$$dc_i/dt = X_i(c_i, \ldots c_k)$$

He then extends the model by taking into account its spatial extension, introducing a system of coordinates (x) on the domain U occupied by the system:

$$\partial_{c_i}(x_i, t)/\partial_t = X_i(c_i, x, t) + k\Delta c_i$$

where the Laplacian $\Delta\ c_i$, assumed to be small with respect to X_i, can be neglected, in Thom's opinion, for structural stability considerations.

This first introduction of Thom's model uses reaction–diffusion equations which, before Thom decided to neglect the Laplacian, are analogous to those of the model presented by Turing in "The Chemical Basis of Morphogenesis" of 1952 [20]. I'm not aware of Thom's further consideration on this first model of reaction–diffusion equations. However, it is remarkable to notice the analogy with Turing's system. Also notably, the two authors also share the same references in theoretical biology (Thompson, Child, Waddington). After having considered, in the next Sect. 12.4 the divergences between Thom and Waddington on Thom's model, we will come back in Sect. 12.5 to a comparison of Thom's and Turing's model with respect the capacity to model the property of developmental canalization.

12.4 Structural Stability and the Epigenetic Landscape

For the embryologist Conrad Hal Waddington, the epigenetic landscape is an answer, in terms of mental images, to his questions about a possible theorisation of embryonic development.

A first image of an epigenetic landscape was introduced by Waddington in *Organisers and Genes* in 1940 ([23], frontispiece). It was taken up and articulated in several images during the 1950s, presented in *Strategy of the Genes* ([28], 29 and 36); based on these Waddington's images, Figs. 12.2 and 12.3 present entirely redrawn images of the epigenetic landscape.

In this representation a ball is ready to roll down a hilly surface (Fig. 12.2). Waddington indicates that the path followed by the ball corresponds to the developmental history of a given part of the egg. The underlying part of the landscape represents the system of interactions woven between the genes, a system that defines the shape of the landscape itself, with the different pegs representing the genes (Fig. 12.3). The strings between the genes represent the chemical patterns of the genes. The epigenetic landscape is shaped by the tension of the strings that branch out from this network below the surface of the landscape.

The relationship of catastrophe theory to experience in embryology is not direct. The mediation is offered by the relevance of the mathematical modelling proposed by Thom as a mathematical expression of the images of the epigenetic landscape and the notions and properties they convey. Such relevance should be guaranteed, according to Thom, by the property of structural stability. However, as Waddington points out to him in their famous correspondence following Thom's submission of "Une théorie dynamique de la morphogenèse" as a part of the first volume of *Towards*

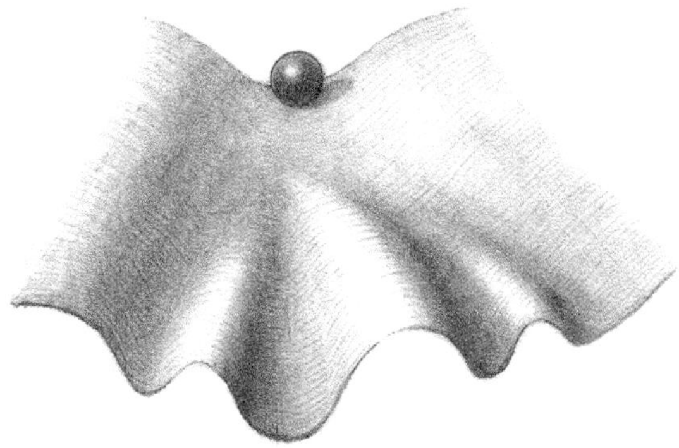

Fig. 12.2 A representation of a part of the epigenetic landscape, in the form of a hilly landscape on which the path of a ball unfolds from the top to the bottom. The path followed by the ball corresponds to the developmental history of a given region of the egg. This is an entirely redrawn image (© Sara Franceschelli) based on the image of the epigenetic landscape in C.H. Waddington, *The Strategy of the Genes*, London, Allen & Unwin, 1957, p. 29

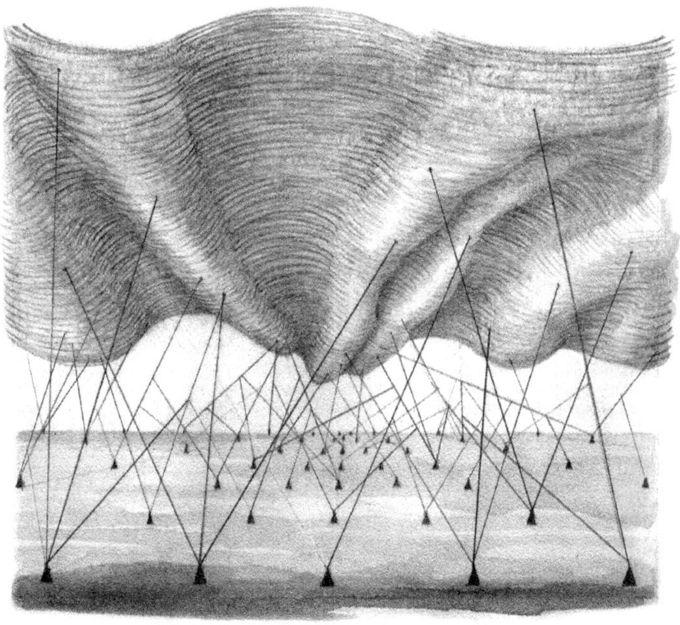

Fig. 12.3 The network of interactions underlying the epigenetic surface. This is an entirely redrawn image, © Sara Franceschelli, based on the image of the network of interactions underlying the epigenetic landscape in C.H. Waddington, *The Strategy of the Genes*, London, Allen & Unwin, 1957, p.36. From the original caption: "The pegs in the ground represent genes; the strings leading from them the chemical tendencies which the genes produce. The modelling of the epigenetic landscape, which slopes down from above one's head towards the distance, is controlled by the pull of these numerous guy-ropes which are ultimately anchored to the genes". *Ibidem*

a Theoretical Biology directed by Waddington,[4] it does not seem to be able to capture all the properties of stability of the developmental processes of a living organism that the epigenetic landscape intends to express.

The correspondence testifies a sort of convergence between the two scholars on the mathematical expression of the notion of "chreod" in terms of "an attractor of biochemical kinetics tangent to the point under consideration" ([16, 18][33]), but it also highlights divergences around the notions of stability that can be used in relation to the landscape and the dynamics characterising it.

Waddington uses his neologism '*homeorhesis*' to designate the property of stability of developmental processes with respect to disturbances in the course of the same processes. To him it is essential not to confuse this property of equilibrium along a chreod with the property of homeostasis, which in his view is only appropriate to designate an equilibrium around a stationary final state. Waddington's *homeorhesis* has a dynamic meaning. The slope of the valleys in the landscape ensures that the ball does not escape the channel, even if it follows a perturbated path—unless the perturbation is not too great, in which case the ball would jump into another valley (for Waddington development is canalized). According to Waddington, Delbrück's model concerns homeostatic dynamics, which tend towards the same final state of equilibrium, not homeorhetic dynamics, in which the equilibrium varies at each moment along a trajectory.

Now, Thom built his theory of catastrophes on the property of structural stability which, being a property of any observable form, allows him to link his vision of morphogenesis to Waddington's landscape. For Thom a given chreod would be defined by a certain region of the parameter space for which a given process is structurally stable. From this definition of the notion of chreod the property of *homeorhesis* would follow naturally. Thom therefore believes that Waddington's reluctance to acknowledge that his system of equations of Delbrück's model amounts to describing a chreod, was due to a difficulty in fully appreciating the mathematical theory in question. Beyond some communication difficulties inherent to the cultural differences between the two scholars, which certainly exist, the obstacles in the search for a complete mathematical expression of the landscape images, which according to Waddington cannot be the same as the one valid for Delbrück's model, is ultimately due, for Waddington, to the different meaning given to the variable time by a mathematician and a biologist:

> Delbrück in 1949 was talking about the alternatives of driving round and round the Place de la Concorde, or round and round the Étoile; and that is only a degenerate case of what I had been talking about in 1940, which is the alternative of taking the bus from the Aérogare des Invalides to the Aéroport Orly or the Aéroport Le Bourget. The only way to eliminate this difference between Delbrück and myself would be if you are so "pure" a mathematician that you acknowledge no difference between a dimension devoted to a material variable, such as concentration, and one devoted to time. But this is a level of abstractness at which mathematics loses touch with the real-world problems of biologists ([16, 18], 34–35)

[4] Partly published as an appendix to the *leading* article on catastrophe theory, "A Dynamic Theory of Morphogenesis" ([16, 18]). Five letters, from 25 January to 27 February 1967. For a detailed study of this correspondence see Franceschelli [8].

12.5 Modelling Canalization? Turing, Thom, and a Still Open Challenge

Thom and Turing also share the same references in biology. Moreover, Waddington was not only René Thom's privileged interlocutor in biology: he was also one of the first biologists to take an interest in Turing's work. In particular, he noted that, despite the title of the article, what Turing's model describes is not really a mechanism of morphogenesis, but a mechanism of pattern formation. Waddington's main criticism of Turing's model is similar to his comments on Thom's model in the seminal article. It highlights specificities of living processes that escape mathematical modelling.

Like other approaches in theoretical biology at the time (e.g. Turing's work on morphogenesis), Waddington's work revives, from the point of view of recognizing the essential contribution of mathematisation to the study of morphogenesis, the perspective developed by Wenworth D'Arcy Thompson in *On Growth and Form* [19].

Thompson expresses a criticism of thinking about morphogenesis on purely adaptationist grounds and the conviction that the challenge is to understand what is shared by living and non-living processes through a mathematical and mechanical approach.

Although Waddington can be considered one of Thompson's heirs, he has never stopped considering himself a Darwinian and an evolutionist. In contrast to Thompson's approach, in the theoretical proposition conveyed by Waddington's images the evolutionary dimension is always present, via the hidden part of the epigenetic landscape, and the effect of genes on embryonic development (and thus on the genesis of forms) is seen as playing a central role. However, contrary to the prevailing view of his time, Waddington did not think in linear terms about the action of an individual gene,[5] but instead took a holistic view, considering the effect of the interaction between all the genes in the organism and their products on development. In a 1942 paper, he uses the term 'epigenetics' to refer to the study of the processes involved in the 'causal mechanisms' by which genotype genes lead to phenotype effects, and defines 'epigenotype' as the 'complex of developmental processes' that link genotype and phenotype ([24], 18). In *Organisers and Genes* of 1940 ([23], 69), Waddington considered that reactions between genes should be represented as non-linear. This position would be maintained for decades to come and would be developed in *The Strategy of the Genes* of 1957. This holistic posture was inspired by the philosophy of Alfred North Whitehead, which he studied during the last two years of his undergraduate degree and which had a great influence on his work from the beginning. Waddington could perhaps be called a 'crypto-metaphysician' since, as he himself writes in an autobiographical note:

> Since I am an unaggressive character, and was living in an aggressively anti-metaphysical period, I chose not to expound publicly these philosophical views ([27], 10)

---

[5] Fox-Keller [6] (ch. I. "Language and Science: Genetics, Embryology, and the Discourse of Gene Action") and Fox-Keller [7] (ch. IV. "Genes, Gene Action, and Genetic Programs"). For a review of Waddington's work see also Hall and Laubichler [1].

In terms of scientific practice, Waddington draws from his reading of Whitehead the idea that instead of taking 'things' as the object of study, we should be interested in 'processes'. Whitehead's processes have an individual character that is highly relational, a 'concrescence' being a unit into which many relationships with other processes converge. It was this idea that led him to develop the images of the epigenetic landscape. I quote here from his intellectual autobiography, discussing how Thom's ideas have, for Waddington, answered questions of Whiteheadian origin:

> In the late thirties I began developing the Whiteheadian notion that the process of becoming (say) a nerve cell should be regarded as the result of the activities of a large number of genes, which interact together to form a unified 'concrescence'. [...] it had become apparent that the 'gene-concrescence' itself undergoes processes of change; at one embryonic period a given concrescence is in a phase of 'competence' and may be switched into one or other of a small number of alternative pathways of further change—but the competence later disappears and if you've missed the bus the switch won't work. My main preoccupation, however, was with the nature of the switches, which was the subject of the experimental biochemical work I was doing, with Needham and others, on 'embryonic induction'. Influenced—probably over-influenced-by genetics, I insisted that the switch must have sufficient specificity to recognize particular genes. We showed that, in these terms, the specificity resides inside the cells which react to induction—we called it 'the masked evocator'. This is very similar to the situation discovered by Jacob and Monod many years later in bacteria, where again the specific repressor molecules are internal to the cells which react to enzyme-inducing substances. If I had been more consistently Whiteheadian, I would probably have realized that the 'specificity' involved does not need to lie in the switch at all, but may be a property of the 'concrescence' and the ways in which it can change. Because of course what I have been calling by the Whiteheadian terms of "concrescence" is what I have later called *a chreod*, a notion which René Thom has explicated; and the switches are Thom's catastrophes. The specificity *need* not be in what precipitates the catastrophe, but could reside only in the possible stable regimes (limit cycle in the simplest case) into which the system could be flipped ([27], 9–10)

The work on embryonic induction and "evocators" referred to above by Waddington is, alongside the form of the reaction–diffusion equations, another element for thinking about the relationship between Turing's and Thom's approach. Turing refers very specifically to Waddington in his work, indicating that Waddington's evocators are an example of morphogens.

In collaboration with Joseph Needham, Waddington was working during the 1930s on the extension of Hans Spemann's theory of embryonic induction and the role in induction played by the organiser (a region of the embryo that produces induction on other, nearby regions).[6] Waddington argued that the outcome of each induction process depends on both the properties of the inducing agent and the material undergoing induction. He thus defined a material capable of reacting to a given inductive stimulus as 'competent' for a given induction process ([21], 179). He thought of a compliant tissue as an unstable system, with different possibilities of evolution, the choice of the path actually chosen depending on the organiser for that tissue ([22], 86).

---

[6] For an analysis of Waddington's work on induction, in the context of the research of the time, at the interface of embryology and genetics, and for the importance of this work on the genesis of the landscape figure, see Gilbert [11].

In *Organisers and Genes* Waddington writes:

> We had described competence as a state of disequilibrium in a complex system of reactants, and had suggested that the reactants are ultimately genes or gene-products ([23], 92)

Waddington considered the action of the organiser, that set of cells capable of inducing cell differentiation, as equivalent to that of genes. He had thus arrived at a conceptualisation of the processes of embryonic development that did not depend on the biochemical description of the substrate.

Several images in *Organisers and Genes* have a similar structure, consisting of paths that at some point branch off. They are used to report either embryology experiments on the organiser, for example in the Amphibia ([23], 6), or genetics experiments on the causal effects of gene inactivation or activation resulting in to different forms of the antenna in Drosophila ([23], 83).

For Waddington what mattered was competence, seen as an unstable equilibrium point, which can respond to both external and internal genetic-type signals. It is this property of instability of some equilibria, which come to mark out flows along a particular path, that the branch points of the landscape seem to indicate:

> It would be difficult to find a similar configuration in any actual piece of country. As one goes downhill, the valley which was originally wide and gently sloping, branches into more and more subdivisions, some of which (representing tracks realised only under the influence of special genes or environmental conditions) may be hanging valleys whose floors disembouch up the side, above the main valley bottom. It is an amusing landscape to picture oneself; and I think it expresses, formally at least, some characteristics of development which are not easy to grasp in any other way ([23], 93)

Given the centrality of stability arguments (understood in a broad sense that also includes the conditions for the establishment of instabilities) in Waddington's research, which is strongly rooted in experimental practice, it is easy to understand the relevance and theoretical scope of Thom's approach, but also of Turing's, for the study of morphogenesis. Despite their differences, in fact, both offer a point of view where changes in the stability of a system, or a family of systems, play a key role. Thus Waddington, in addition to recognising the relevance of Thom's mathematical translation of the notion of the chreod, recognises in Thom's catastrophes the "switches" between chreods. On the other hand, Turing's work had already highlighted that the theoretical heart of his model lay in the study of the establishment of instabilities in the reactions between morphogens, thus allowing access to a theoretical level of modelling of morphogenetic processes irreducible to physicalist materialism. In this respect, it will suffice to recall the relational definition given by Turing to morphogens, the actors of his reaction–diffusion equations:

> These substances will be called morphogens, the word being intended to convey the idea of a form producer. [...] "[the term "morphogen"] is not intended to have any exact meaning, but it is simply the kind of substance concerned in this theory" ([20], 38)

A morphogen is therefore a chemical substance that enters into a certain relationship with another chemical substance (another morphogen), a relationship that can be described by reaction–diffusion equations. It is the study of the appearance of instability that allows us to understand the formation of a pattern compared to the previous homogeneous situation. In this sense the definition of morphogen is relational. It follows that different substances behaving in interaction as defined by the Turing system can potentially be considered as morphogens.

In fact, Waddington expresses his interest in Turing's approach in a September 1952 letter.[7] He believes, however, that the kind of process described by Turing plays a role in the formation of patterns, such as the pelage spots of animals, the patterns on the wings of butterflies or on the shells of molluscs, but he doubts that it plays a very important role in the morphogenesis occurring in the early stages of development. Waddington believes that the newly fertilised egg does not satisfy the homogeneity condition that Turing's model assumes, but instead has an initial pattern element of its own.

Waddington details in his letter the *requirements* that mathematical models adapted to describe the early stages of embryological development should meet.

> The problem of embryological interest which I should most like to see tackled from a mathematical point of view is the following: *Development is particularly characterised by the fact that it produces a finite number of quite definitely distinct tissues and organs and does not produce all intermediate types of tissues between the kidney and the liver for instance*. If one imagines a series of synthetic chemical processes, probably autocatalytic and interfering with or stimulating one another, for instance by competing for substrates or in other ways, *under what conditions will the system have a finite number of distinct paths, which it may follow?* What sort of alterations would be necessary to cause the development to click over from one path into another alternative?[8]

While Waddington expresses interest in the kind of autocatalytic models that Turing proposes, he believes that Turing's model cannot represent a process with a finite number of distinct alternative pathways, without the possibility of intermediate solutions. This comment by Waddington provides a privileged access to understanding the difficulties of translating concepts describing biological processes into mathematical terms. Although Waddington does not use the term, what is at issue here is the possibility of expressing mathematically the notion of 'canalization', which Waddington introduced in 1942. Since its introduction, this notion has been deeply intertwined with arguments from natural selection and genetics. In his article "Canalization of development and the inheritance of acquired characters" [25] Waddington introduces canalization as a property of developmental reactions that depends on the nature of the same reactions. It is because these developmental reactions occur in organisms subject to natural selection that they are, for Waddington, canalized.

> That is to say, they are adjusted so as to bring about one definite end-result regardless of minor variations in conditions during the course of the reaction ([25], 563)

From this point of view, the neologism 'canalization' seems to indicate the property of 'robustness', which in the contemporary language of complex systems refers

[7] Letter from Waddington to Turing, 11 September 1952, from Turing Digital Archive, AMT/D/1TLS (images 19–20).

[8] *Ibidem*, italics mine.

to the ability of a system to withstand different types of disturbance. The term 'robustness' is not used by Waddington, however.

It is evident that the impasse in the mathematical modelling of the canalization property is not only a matter for the Turing model. In his 1989 inventory of Waddingtonian terms, Thom points out that the structural stability property is unable to express canalization, since canalization has a local and metric character, unlike structural stability, which has a global topological character[9]:

> [Canalization] describes any kind of process whose temporal evolution is buffered against external perturbations. There is apparently no strict mathematical equivalent to this concept, as the classical 'structural stability' is of global topological nature, whereas 'canalization' has a metric and local character ([15], 3)

Thom's remark could be transformed into an invitation, for those who would like to follow the path opened up by his morphological work, to push further the questioning of the mathematisation of this type of living process. From this point of view, the notion of canalization is promising because, like that of "genetic assimilation" [26], it is constantly related in Waddington's writings to the dynamic and plastic properties of the landscape in relation to internal or external stimuli.

Several decades after their formulation, the notions of canalization and genetic assimilation were successfully reconsidered and expressed in terms of molecular biology, starting with the work of Rutherford and Lindquist on the "*heatshock protein*" HSP90 in Drosophila development [13]. This is a belated recognition of the theoretical value of Waddington's holistic and non-reductionist work, which can only rekindle interest in Thom's work, which provided the mathematical tools to think in terms of structural stability, attractors and changes in attractors, Waddington's landscape. It would thus be a matter of taking into account a connotation of Waddington's landscapes that is not discussed in Thom's first texts on catastrophe theory, but which subsequently attracted his attention, particularly during the development of his work on saliencies and pregnances: their plastic connotation, i.e., their ability to retain a form imparted by the exterior. It is a connotation that, for Waddington, is closely linked to a Darwinian and population-based interpretation of landscapes. In a 1959 paper, Waddington states that the course of normal development can be represented by the trough of a valley, the sides of which represent the opposition that the system presents to stresses that could disrupt it; more precisely, a section of the valley, i.e., a chreod, represents the system's adaptability curve ([29], 392).[10] It is at the meeting of this population, statistical and Darwinian reading of the landscape, which has already had its recognition in molecular biology, that the Thomian programme could be, paradoxically, pursued.

---

[9] I thank Peter Saunders for an interesting conversation on this question.

[10] I am grateful to Jean Gayon(+) for drawing my attention to the population connotation of Waddington's landscapes. The reader may wish to refer to Gayon [10].

References

1. Andronov, A.A., Pontrjagin, L.: Coarse systems. Dokl. Akad. Nau. S.S.S.R. **14**(5), 247–250 (1937)
2. Andronov, A.A., Vitt, A.A., Khaikin, S.: Theory of Oscillators. Pergamon Press (1966). (Tr. From the Russian, First Edition Moscow, 1937)
3. Chaperon, M.: Qu'est-ce que la stabilité structurelle? Chaos et systèmes dynamiques. In: Franceschelli, S., Roque, T., Paty, M. (eds.) Eléments pour une épistémologie, pp. 79–109. Hermann, Paris (2007)
4. Delbrück, M.: Unités biologiques douées de continuité génétique, CNRS, Paris. Evol. Bull. Hist. Epistem. Sci. Vie **5**(2), 133–166 (1949)
5. Fox-Keller, E.: Refiguring Life-Metaphors of Twentieth-Century Biology. Columbia University Press (1995)
6. Fox-Keller, E.: Making Sense of Life, Explaining Biological Development with Models, Metaphors and Machines. Harvard University Press (2002)
7. Franceschelli, S., Petitot, J.: Commentaires de Sara Franceschelli et Jean Petitot. Œuvres Mathématiques **II**, 343–362 (2019)
8. Franceschelli, S.: Morphogenèse, stabilité structurelle et paysage épigénétique. Morphogenèse. In: Bourgine, P., Lesne, A. (eds.) L'origine des forms, pp. 298–308. Belin, Paris (2006). (Eng. Tr. Morphogenesis. Origins of Patterns and Shapes, Berlin and Heidelberg, Springer Verlag, 2011)
9. Gayon, J.: The marginalization of form in evolutionary biology. Bull. Hist. Epistem. Sci. Vie **5**(2), 133–166 (1998)
10. Gilbert, S.F.: Induction and the Origins of Developmental Genetics. A Conceptual History of Modern Embryology, vol. 7, pp. 181–206. Plenum Press, New York (1991)
11. Hall, B.K., Laubichler, M. (eds.): Conrad Hal Waddington, theoretical biology, and EvoDevo thematic issue. Biol. Theory **3**(3). (Summer 2008)
12. Peixoto, M.: Structural stability on 2-dimensional manifolds. Topology **1**, 101–120 (1962)
13. Rutherford, S., Lindquist, S.: Hsp90 as a capacitor for morphological evolution. Nature **396**(1998), 336–342 (1998)
14. Smale, S.: What is global analysis? In: Smale, S. (ed.) The Mathematics of Time: Essays on Dynamical Systems, Economic Processes, and related Topics, pp. 84–89. Springer, New York (1980)
15. Thom, R.: An inventory of Waddingtonian concepts. In: Goodwin, B., Saunders, P. (eds.) Theoretical Biology. Epigenetic and Evolutionary Order from Complex Systems, pp. 1–7. Edinburgh University Press (1989)
16. Thom, R.: Modèles mathématiques de la morphogenèse, Christian Bourgeois Éditeur, Paris, second revisioned and augmented edition, 315 p (1980)
17. Thom, R.: On the work of Stephen Smale. In: Proceedings of the International Congress of Mathematicians, Moscow, 1966, pp. 25–28. Izdat. Mir, Moscow (1966)
18. Thom, R.: A dynamic theory of morphogenesis. In: Waddington, C.H. (ed.) Towards a Theoretical Biology I, pp. 152–166. University of Edinburgh Press (1968)
19. Thompson, D.A.W.: On Growth and Form, 2nd edn, 1942, reprinted 1945. Cambridge University Press (1917)
20. Turing, A.M.: The chemical basis of morphogenesis. Philos. Trans. Roy. Soc. B **237**, 37–72 (1952)
21. Waddington, C.H.: Experiments on the development of chicken and duck embryos. Philos. Trans. Roy. Soc. B London **221**(1932), 179–230 (1932)
22. Waddington, C.H.: The origin of competence for lens formation in the amphibia. J. Exp. Biol. **48**(1936), 86–91 (1936)
23. Waddington, C.H.: Organisers and Genes. Cambridge University Press (1940)
24. Waddington, C.H.: The epigenotype. Endeavour **1**, 18–20 (1942)
25. Waddington, C.H.: Canalization of development and the inheritance of acquired characters. Nature **3811**(1942), 563–565 (1942)

26. Waddington, C.H.: Genetic assimilation of an acquired character. Evolution **7**, 118–126 (1953)
27. Waddington, C.H.: The Evolution of an Evolutionist. Cornell University Press, Ithaca (1975)
28. Waddington, C.H.: The Strategy of the Genes. Allen & Unwin, London (1957)
29. Waddington, C.H.: Evolutionary adaptation. Perspect. Biol. Med. **2**(4), 379–401 (1959). (Summer 1959)

Part IV
René Thom, a Personal Testimony

Chapter 13
The Thom Seminary: Its Genesis

Claude P. Bruter

Born in 1923, René Thom cheerfully went through all the stages of a brilliant academic career. His mathematical work is in line with that of Russian and American mathematicians. Thom published his first note in the Comptes-Rendus at the age of 26. The content of this note, « *Sur une partition en cellules associée à une fonction sur une variété* » (« *On a partition into cells associated with a function on a manifold* ») foreshadows that of all his later mathematical work. The properties of the function he uses announce his interest in the theory of singularities of differentiable maps. The partition into cells of a manifold refers to its interest in topology, in particular the classification of differentiable manifolds.

He published his thesis two years after this first note in the Comptes-Rendus, and obtained the Fields Medal in 1958, at 35 years of age, for his work in algebraic topology on the classification of differentiable varieties using significant groups. The use of the assembly of varieties by cobordism, that is to say their association according to compatible boundaries, allowed him to widen the field of the results already known.

In one of our conversations, around 1975 I think, he revealed to me spontaneously, blushing, that the mathematician Gaston Julia had advised him to study singularities. I made the mistake of not asking him about when this suggestion came about. He worked in this direction as an extension of the work of Marston Morse and Hassler Whitney. He once told me that he considered Whitney to be the most important mathematician of his century.[1] This was of course a personal point of view.

I no longer remember how I was informed of a presentation by Thom, at the mathematics department of Orsay, on a Thursday in early June 1968.

[1] One of the points of connection between Thom and me was the fact that Whitney was also the founder of the theory of matroids, the subject of my main thesis.

C. P. Bruter (✉)
Université Paris12, Créteil, France
e-mail: bruter@me.com

© The Author(s), under exclusive license to Springer Nature Switzerland AG 2024
I. Marcos and C. Morier (eds.), *The Relevance of René Thom*, Lecture Notes in Morphogenesis, https://doi.org/10.1007/978-3-031-54982-3_13

It was afternoon, the sky was all blue and the the sun was shining. I was sitting on the left side, facing the desk in front of which Thom was presenting some of his views on biology and linguistics. Thom spoke of this using the elements of his theory of course. The text appeared in number 4, December 1968, of the journal *l'Âge de la Science* (*Age of Science*), under the title *Topologie et Signification* (*Topology and Meaning*), and it specified the content of the part of the presentation devoted to linguistics.

I was fascinated by the cleverness and wealth of this presentation, whose mathematical roots were somewhat magical for me at the time, delighted if not joyful because I encountered in it the geometric attitude of my own approach to linguistics. At the time, I had already written my first article on linguistics entitled *Sur une formalisation de la grammaire française simplifiée*[2] (*On a formalization of simplified French grammar*). I relied on the fact that an important part of speech, through the simulation activity performed by the brain, reflects the spatial organization of objects. I found in Thom's underlying presentation the same point of view.

My article appeared in 1970 only in *Etudes de Linguistique Appliquée*. This delay in publication allowed me to complete the original writing simply by adding Thom's name in the introduction, without giving any bibliographical reference, and in particular concerning him since the approach, of an algebraic nature, was very different from his, which was centered on a form of analytical representation of certain verbs.

The following is the introduction to this text[3]:

SUR UNE FORMALISATION
DE LA GRAMMAIRE FRANÇAISE SIMPLIFIÉE

0. *Introduction.*

Dans l'essai qui suit, on présente une formalisation abrégée de la grammaire française simplifiée, en tenant compte de quelques-unes de ses propriétés sémantiques.

Il nous est apparu que certaines règles fondamentales de la construction de nombreux langages, y compris le langage mathématique, rentraient dans le cadre de ce formalisme. Il présente donc un intérêt pédagogique et méthodologique.

Son choix repose sur le fait qu'un langage a pour but de décrire les objets dans l'espace et dans le temps, ainsi que leurs interactions et mouvements. Ce fait, pressenti par N. Chomsky, semble assez important pour avoir, indépendamment de nous, retenu l'attention de R. Thom, et lui permettre d'introduire de nouveaux modèles mathématiques en linguistique.

[2] Accessible on http://arpam.free.fr/FGFS.pdf.

[3] On a formalization of simplified French grammar. In the following essay, we present an abbreviated formalization of simplified French grammar, taking into account some of its semantic properties. We learned that certain fundamental rules for the construction of many languages, including mathematical language, fell within the scope of this formalism. It is therefore of pedagogical and methodological interest. Its choice is based on the fact that a language aims to describe objects in space and time, as well as their interactions and movements. This fact, sensed by N. Chomsky, seems important enough to have, independently of us, caught the attention of R. Thom, and allowed him to introduce new mathematical and linguistic models.

So sharing with Thom the same spatio-temporal understanding of language, I came to see him at the end of his presentation, told him of my feeling of complicity with his geometric vision of language. He invited me to come and meet him at 11 a.m. at IHES, the last Saturday of the month.

That day, we resumed an appointment for the Saturday of the new school year, again at 11 a.m., at the IHES. We were thus alone for at least two years, after which other people joined, the first of them being Pierre Delattre†. Many who attended what became a regular seminar are present here, but there were also many other participants, more or less episodic: Delattre's two collaborators at the CEA, Hyver and Delforge, my late friends, the linguist Maurice Coyaud[4] (†), who knew almost all the languages and dialects of Asia!, the mathematician Jean-Pierre Duport-Rosand[5] (†), author of a very beautiful tale *La princesse bleue* (*The Blue Princess*), inspired by the notion of chreod, and author of a great application of CT to hemophilia, the philosopher of science Jean Largeault[6] (†) of course, the morphologist, biologist and physicist Yves Bouligand[7] (†), the philosopher Kristof Pomian, the international lawyer Alain Laraby, the neurobiologist Pierre Etevenon, whose guinea pig I was for a while, and several others whose names you know, who are now absent.

Thom remained practically silent during our one-on-ones sessions where a relaxed atmosphere of perfect understanding reigned between us. We didn't talk about biology in any way. I explained my ideas to him, most of which, written at the CNRS center in Aussois in September, was the subject of my mimeographed text of October–November 1969 entitled *On the cave wall* (*Sur le Mur de la Caverne*[8]). The content of this essay provided the material, and the first part of Volume 1, and Volume 3 of *Topologie et Perception*[9] (*Topology and Perception*). In the essay, the announcement of the expected publication of *Stabilité Structurelle and Morphogenèse* (*Structural Stability and Morphogenesis*) appears in the bibliography. The book was to appear in 1970. It was finally published in November 1972, and André Avez had guessed perfectly why the last chapters bear the discreet traces of my comments. They are not very visible, even to me now, because Thom had integrated them into his vision of things at the time. I would not be surprised if the Appendix on the notion of object was added following our meetings. Thom had only given me the proofs of these last chapters. We can see, in the preface he wrote for *Topologie et Perception,* the expression of our intellectual sympathy and no doubt a certain form of acknowledgment of debt.

What I learned from Thom therefore only comes, as for everyone, from reading some of his texts and from his public presentations. In particular, I wrote up his lessons given on Monday afternoons at the IHES throughout the 1970–1971 academic year,

[4] Directeur de Recherches CNRS, Prof à l'INALCO.

[5] Maître de Conférences à l'Université de Savoie.

[6] Professeur à l'Université Paris 12.

[7] Président après Thom de la Société Française de Biologie Théorique.

[8] http://arpam.free.fr/SMC.pdf.

[9] http://arpam.free.fr/DP.pdf.

and entitled *Modèles mathématiques de la Morphogenèse*[10] (*Mathematical Models of Morphogenesis*). Thom told me he would thank me, but… These texts do not appear in the works of Thom published by mathematicians. They served as the basis for the publication of the *Lezione Fermiane* included in his book edited by Flammarion. The introduction where the problem of explanation appears, for example, is mine, there is also the use of the term extremality, still unknown to dictionaries. I did not bring anything on the mathematical level, I learned it. I did, however, contribute to the formulation of his model. The first two chapters were not modified. Only the second part of the third chapter had been reviewed by Thom, as at the end of page 32: « I gave … ».

What I have retained, in an essential way, from this long contact with the writings of Thom, is the importance of the notion of singularity (introduced by Cauchy), and that of stability. Let me quote a few lines from the "Advice to the Reader" that appears in my book titled *Energie et Stabilité*[11] and Stability:

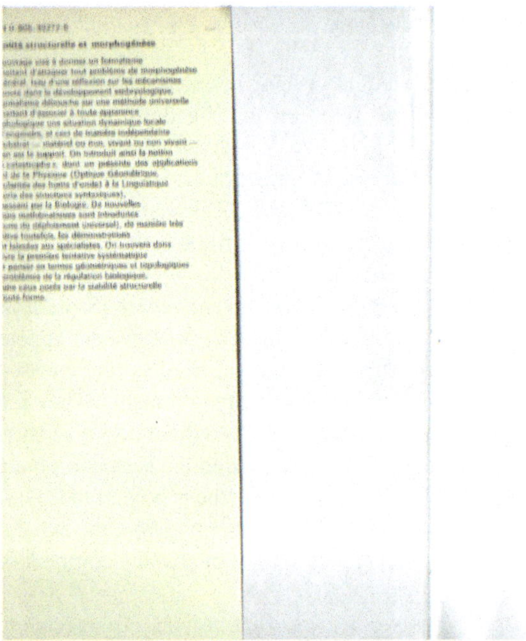

A word about the focus in this text on stability. It comes from my meetings with Thom. First of all, I am grateful to him for having directed me to André Avez, who died recently, quietly regarding the mathematical community. I would like to pay homage to him here. By assigning me the subject of my second thesis, Avez introduced me to the theory of dynamical systems; the notion of stability occupies an eminent place. At that time, influenced by the popularity of structuralism, the

[10] http://arpam.free.fr/MMM.pdf.

[11] Accessible on http://arpam.free.fr/ESC.pdf.

notion of structural stability occupied a prominent place in people's minds. Thom, in his writings, speaks essentially only of this one. I felt the need to go beyond this restrictive point of view. I add that this was all the more necessary since structural stability is by no means generic, as Smale has shown.

I have made this notion of stability a sort of operational law, analogous to certain fundamental laws of physics, such as those of gravitation or those of Maxwell, a law of a metaphysical nature given the observed universality of its field of application. I call it "Plato's principle" (cf T*he Symposium,* 207d) and state it today in this form: *Every object strives to persevere in its "self", across space and time.* If Pascal had read Plato's Banquet, he would have found the answer to one of his questions. Taking up and developing the content of the first part of *Topology and Perception* from the angle of the Mother Ideas,[12] a book nearing completion reveals the extent of the impregnation of the principle and its driving force in all of the activities of the natural world.[13]

---

[12] The Principe of Stability, an Essay on the Incarnation of the Notion of Stability within the Pantheon of Mother Ideas Cambridgescholars.com, 2021 https://www.cambridgescholars.com/product/978-1-5275-6857-0.

[13] I would like to thank Sharon Breit-Giraud who generously corrected my first translation of the original French text.

SPRINGER NATURE

GPSR Compliance

The European Union's (EU) General Product Safety Regulation (GPSR) is a set of rules that requires consumer products to be safe and our obligations to ensure this.

If you have any concerns about our products, you can contact us on ProductSafety@springernature.com

In case Publisher is established outside the EU, the EU authorized representative is:

Springer Nature Customer Service Center GmbH
Europaplatz 3
69115 Heidelberg, Germany

The manufacturer's authorised representative in the EU is Springer Nature Customer Service Centre GmbH, Europaplatz 3, 69115 Heidelberg, Germany. If you have any concerns regarding our products, please contact ProductSafety@springernature.com

Printed and bound by CPI Group (UK) Ltd, Croydon, CR0 4YY

29/07/2026

02180492-0001